Fuzzy Management Methods

More information about this series at http://www.springer.com/series/11223

Andreas Meier • Edy Portmann • Luis Terán
Editors

Applying Fuzzy Logic for the Digital Economy and Society

Springer

Editors
Andreas Meier
Department of Informatics
University of Fribourg
Fribourg, Switzerland

Edy Portmann
Human-IST Institute
University of Fribourg
Fribourg, Switzerland

Luis Terán
Department of Computer Science
Universidad de las Fuerzas Armadas
(ESPE)
Sangolquí, Ecuador

Department of Informatics
University of Fribourg
Fribourg, Switzerland

ISSN 2196-4130 ISSN 2196-4149 (electronic)
Fuzzy Management Methods
ISBN 978-3-030-03367-5 ISBN 978-3-030-03368-2 (eBook)
https://doi.org/10.1007/978-3-030-03368-2

Library of Congress Control Number: 2019931371

This Springer imprint is published by the registered company Springer Nature Switzerland AG
The registered company address is: Gewerbestrasse 11, 6330 Cham, Switzerland

Preface

With today's information overload, it has become increasingly difficult to analyze the huge amounts of data and to generate appropriate management decisions. Furthermore, the data are often imprecise and will include both quantitative and qualitative elements. For these reasons, it is important to extend traditional decision-making processes by adding intuitive reasoning, human subjectivity, and imprecision.

In the age of big data, decision-making processes for economy and society have to deal with uncertainty, vagueness, and imprecision. Besides volume, variety, and velocity, two others V's for veracity and value have also to be taken into consideration. Therefore, the application of fuzzy sets and fuzzy logic becomes a hot topic.

In 2008, the Department for Informatics at the University of Fribourg, Switzerland, founded its Research Center for Fuzzy Management Methods (FMM = FM^2), often only called FMsquare. Later on, the International Research Book Series for FMsquare was launched by Springer, where researchers published in fuzzy-based reputation management, fuzzy classification of online customers, inductive fuzzy classification for marketing analytics, fuzzy data warehousing for performance measurement, using intuitionistic fuzzy sets for service level engineering, building a knowledge carrier based on granular computing, and a fuzzy-based recommender system for political communities, among others.

To celebrate the 10th anniversary of FMsquare in 2018, international researchers have been invited to submit their contributions in the following topics:

- Fuzzy-based portfolio management
- Web analytics with fuzzy measures
- Community marketing with fuzzy approaches
- Fuzzy-based customer equity
- Business process modeling with words
- Data mining with fuzzy reasoning
- Fuzzy cognitive maps for knowledge management
- Fuzzy-based stakeholder management

- Sense-making with vague data
- Related topics

After the international call for book chapters in November 2017, ten chapters were selected for this book in April 2018. The book presents state-of-the-art methods, case studies, and web-based services for a digital economy and society. The target audience are researchers, practitioners, project leaders, politicians, and managers who like to apply or improve fuzzy-based skills.

Fribourg, Switzerland Andreas Meier
Fribourg, Switzerland Edy Portmann
Sangolquí, Ecuador Luis Terán
July 2018

Contents

Contributors

Jose M. Alonso Centro Singular de Investigación en Tecnoloxías da Información (CiTIUS), Universidade de Santiago de Compostela, Santiago de Compostela, Spain

Ahmad Zaki Mohamad Amin Mathematics Division, Centre for Foundation Studies in Science, University of Malaya, Kuala Lumpur, Malaysia

Ahmad Syafadhli Abu Bakar Mathematics Division, Centre for Foundation Studies in Science, University of Malaya, Kuala Lumpur, Malaysia

Centre of Research for Computational Sciences and Informatics in Biology, Bioindustry, Environment, Agriculture and Healthcare (CRYSTAL), University of Malaya, Kuala Lumpur, Malaysia

Rédina Berkachy Applied Statistics and Modelling, Department of Informatics, Faculty of Economics and Social Sciences, University of Fribourg, Fribourg, Switzerland

Roberto Evelim Penha Borges Universidade Federal do Rio Grande do Norte, Natal, Brazil

Adrian Castro-Lopez Business Administration Department, Faculty of Economic and Business, University of Oviedo, Oviedo, Spain

Marco Antonio Guimarães Dias Pontifícia Universidade Católica do Rio de Janeiro, Rio de Janeiro, Brazil

Laurent Donzé Applied Statistics and Modelling, Department of Informatics, Faculty of Economics and Social Sciences, University of Fribourg, Fribourg, Switzerland

Ivana Dragović Faculty of Organizational Sciences, University of Belgrade, Belgrade, Serbia

Alexander Gegov School of Computing, University of Portsmouth, Portsmouth, UK

Aleksandar Janjić Faculty of Electronic Engineering, University of Niš, Niš, Serbia

Ku Muhammad Naim Ku Khalif Department of Science Program (Mathematics), Faculty of Industrial Sciences and Technology, Universiti Malaysia Pahang, Pahang, Malaysia

José Mancera Department of Informatics, University of Fribourg, Fribourg, Switzerland

Andreas Meier Université de Fribourg, Fribourg, Switzerland

Jaime Meza Universidad Técnica de Manabí (UTM), Portoviejo, Ecuador

Pavle Milošević Faculty of Organizational Sciences, University of Belgrade, Belgrade, Serbia

Adrião Duarte Dória Neto Universidade Federal do Rio Grande do Norte, Natal, Brazil

Minh Tue Nguyen Human-IST Institute, University of Fribourg, Fribourg, Switzerland

Mayank Pandey Department of Electrical Engineering, Indian Institute of Technology Kanpur, Kanpur, India

Bratislav Petrović Faculty of Organizational Sciences, University of Belgrade, Belgrade, Serbia

Ana Poledica Faculty of Organizational Sciences, University of Belgrade, Belgrade, Serbia

Edy Portmann Human-IST Institute, University of Fribourg, Fribourg, Switzerland

Aleksandar Rakićević Faculty of Organizational Sciences, University of Belgrade, Belgrade, Serbia

Vikas Singh Department of Electrical Engineering, Indian Institute of Technology Kanpur, Kanpur, India

Luis Terán University of Fribourg, Fribourg, Switzerland

Universidad de las Fuerzas Armadas ESPE, Sangolquí, Ecuador

Martha Tomalá Municipality of District of Quito (MDMQ), Quito, Ecuador

Jelena D. Velimirović Mathematical Institute of the Serbian Academy of Sciences and Arts, Belgrade, Serbia

Lazar Z. Velimirović Mathematical Institute of the Serbian Academy of Sciences and Arts, Belgrade, Serbia

Nishchal K. Verma Department of Electrical Engineering, Indian Institute of Technology Kanpur, Kanpur, India

Abdul Malek Yaakob School of Quantitative Sciences, Universiti Utara Malaysia, Kedah, Malaysia

Chapter 1
Testing Hypotheses by Fuzzy Methods: A Comparison with the Classical Approach

Rédina Berkachy and Laurent Donzé

1.1 Introduction and Motivation

Nowadays, big data is challenging computer scientists as well as statisticians. Indeed, treating large and complex data sets from multiple sources has become a necessity. Applications can be found in the areas of management, processing, analysis, and others. The so-called digital economy is undoubtedly influencing the way statisticians are accomplishing their tasks. Different aspects of the digital economy are reflected in the data we are collecting. Since human opinion is in many cases uncertain, convenient statistical tools monitoring these aspects should be well adapted. When data contains fuzziness, extracting precise information becomes much more difficult. Fuzzy logic was introduced for dealing with such problems. The extension of statistical methods from the classical approach to the fuzzy context has not yet been fully revealed. We note for instance the hypotheses testing statement and particularly the computation of p-values.

Inspired by Grzegorzewski [6] who considered data as fuzzy, an approach of testing hypotheses by confidence intervals in the fuzzy environment was shown by Berkachy and Donzé [4]. To complement the previous work and based on Filzmoser and Viertl [5] and Parchami et al. [7], a fuzzy p-value and its α-cuts were described by Berkachy and Donzé [3]. We recall that Filzmoser and Viertl [5] assumed that fuzziness basically comes from the data, and contrariwise, Parchami et al. [7] considered that fuzziness is in the hypotheses.

Our first contribution is to show an approach where we consider that both the data and the hypotheses are fuzzy. In addition, it is evident to see that the resulting decisions of such approaches are fuzzy and therefore in many situations

R. Berkachy (✉) · L. Donzé
Applied Statistics and Modelling, Department of Informatics, Faculty of Economics and Social Sciences, University of Fribourg, Fribourg, Switzerland
e-mail: Redina.Berkachy@unifr.ch; Laurent.Donze@unifr.ch

A. Meier et al. (eds.), *Applying Fuzzy Logic for the Digital Economy and Society*, Fuzzy Management Methods, https://doi.org/10.1007/978-3-030-03368-2_1

can be difficult to interpret. In such situations, defuzzification is often required. We advocate the signed distance defuzzification method and propose our defuzzified approach of testing with its corresponding decision rules. One of the principal contributions is to show how one can use our methods with real life data. Toward this aim, we apply our procedures on a survey from a financial institution of Zurich. The purpose is to analyse our data set using fuzzy approaches of hypotheses testing in order to show their usability.

We then perform the same tests with the classical approach. Both approaches are then compared in order to reveal the differences between them. The last and main objective of this chapter is to provide clear guidelines and recommendations for the use of both methods in the purpose of being rigorously used. One of our major results is that in many cases both approaches give similar decisions related to rejecting or accepting a given hypothesis. We should be careful in interpreting the decisions obtained from the fuzzy approaches in particular when observing their assigned degrees of conviction.

We present in Sect. 1.2 some useful definitions and notations. Section 1.3 is devoted for the testing hypotheses statement by the classical approach. In Sect. 1.4, we outline the testing hypotheses statement by the fuzzy approach where a procedure with confidence intervals and fuzzy p-values is described. In Sect. 1.5, we show how to defuzzify the obtained test decisions by the signed distance. Our approaches are illustrated by numerical examples from real data in Sect. 1.6, and the differences between the related decisions made by the classical and the fuzzy approaches are discussed. We close the chapter with some guidelines and recommendations for using both approaches.

1.2 Definitions and Notations

Definition 1.1 (Fuzzy Set) If A is a collection of objects denoted generically by x, then a fuzzy set $\tilde{X}$ in A is a set of ordered pairs:

$$\tilde{X} = \{(x, \mu_{\tilde{X}}(x)) : x \in A\}, \tag{1.1}$$

where $\mu_{\tilde{X}}$(x) is the membership function of x in $\tilde{X}$, which maps A to the closed interval [0,1] that characterizes the degree of membership of x in $\tilde{X}$.

Definition 1.2 (Fuzzy Number) A fuzzy number $\tilde{X}$ is a convex and normalized fuzzy set on $\mathbb{R}$, such that its membership function is continuous and its support is bounded.

Definition 1.3 (α-Cut of a Fuzzy Number) The α-cut of a fuzzy number $\tilde{X}$ is a non-fuzzy set defined as:

$$\tilde{X}_\alpha = \{x \in \mathbb{R} : \mu_{\tilde{X}}(x) \geqslant \alpha\}. \tag{1.2}$$

The fuzzy number $\tilde{X}$ can be represented by the family set $\{\tilde{X}_\alpha : \alpha \in [0, 1]\}$ of its α-cuts.

The α-cut of a fuzzy number $\tilde{X}$ is the closed interval $[\tilde{X}^L_\alpha, \tilde{X}^R_\alpha]$, for which $\tilde{X}^L_\alpha$ and $\tilde{X}^R_\alpha$, its left and right α-cuts, are given respectively by:

$$\tilde{X}^L_\alpha = \inf\{x \in \mathbb{R} : \mu_{\tilde{X}}(x) \geqslant \alpha)\} \text{ and } \tilde{X}^R_\alpha = \sup\{x \in \mathbb{R} : \mu_{\tilde{X}}(x) \geqslant \alpha)\}.$$

1.3 Testing Hypotheses by the Classical Approach

We consider a population described by a probability distribution P_θ depending on the parameter θ. It belongs to a family of distributions $\mathbb{P} = \{P_\theta : \theta \in \Theta\}$. For testing hypotheses on a parameter θ by the classical approach, we consider a null hypothesis denoted by H_0, H_0: $\theta \in \Theta_{H_0}$ and an alternative one denoted by H_1, H_1: $\theta \in \Theta_{H_1}$. Θ_{H_0} and Θ_{H_1} are the subsets of Θ such that $\Theta_{H_0} \cap \Theta_{H_1} = \emptyset$. Let $X_1, \ldots, X_n$ be a random sample. We denote by T, a test statistic, i.e., a function of this sample used in testing the null hypothesis against the alternative one, where T: $\mathbb{R}^n \mapsto \mathbb{R}$. In this case, one can make any of these two decisions:

"reject the null hypothesis H_0" or "not reject the null hypothesis H_0".

The decision is made based on the test statistic T. For this purpose, we define a space of possible values of T decomposed into a rejection region R and its complement R^c. The rejection region R can be of three forms depending on the alternative hypotheses H_1.

The following three tests are considered:

1. $H_0 : \theta \geq \theta_0$ vs. $H_1 : \theta < \theta_0$; (1.3)
2. $H_0 : \theta \leq \theta_0$ vs. $H_1 : \theta > \theta_0$; (1.4)
3. $H_0 : \theta = \theta_0$ vs. $H_1 : \theta \neq \theta_0$; (1.5)

where θ is the parameter to test and θ_0 a particular value of this parameter. Based on the statistic T, we would reject the null hypothesis H_0 if respectively:

1. $T \leq t_l$ (one-sided test); (1.6)
2. $T \geq t_r$ (one-sided test); (1.7)
3. $T \notin (t_a, t_b)$ (two-sided test); (1.8)

where t_l, t_r, t_a and t_b are quantiles of the distribution of T.

1.4 Testing Hypotheses by the Fuzzy Approach

Many researchers have treated the hypothesis testing problem with the fuzzy approach. For instance, Grzegorzewski [6] proposed a testing procedure with fuzzy data based on confidence intervals. Inspired by this approach, we presented in Berkachy and Donzé [4] an approach based on confidence intervals, but we asserted that fuzziness comes from both the data and the hypotheses. A decision can be made by calculating a p-value and comparing it to the significance level.

We note that by definition a p-value is the probability of observing a particular sample given that the null hypothesis is true. In this perspective, some researches focused on fuzzy p-values: Filzmoser and Viertl [5] gave the fuzzy p-values by considering that the fuzziness in the tests comes from the data. In the same way, Parchami et al. [7] assumed that fuzziness is a matter of hypotheses. Berkachy and Donzé [3] generalized both previous works and provided a fuzzy p-value asserting that in many cases fuzziness can come from the data or from the hypotheses.

We discuss in the following the approaches proposed by Berkachy and Donzé [4] and [3]. We close the section by defuzzifying the obtained fuzzy decision by the signed distance, as described in these papers. First, let us discuss the concept of fuzzy hypothesis.

Definition 1.4 (Fuzzy Hypothesis) A fuzzy hypothesis $\tilde{H}$ on the parameter θ, denoted as "$\tilde{H} : \theta$ is H", is a fuzzy subset of the parameter space Θ with its corresponding membership function $\mu_{\tilde{H}}$.

Remark 1.1 A given fuzzy hypothesis $\tilde{H}$ is a generalization of the crisp hypothesis H. A crisp hypothesis has a membership function $\mu_{\tilde{H}} = I_{\Theta}$.

Fuzzy hypotheses can be modelled by different shapes of fuzzy numbers. It is common practice to use triangles.[1] With regard to the tests (1.3), (1.4) and (1.5), the fuzzy hypotheses with triangular membership functions are as follows:

1. $\tilde{H}^{OL} = (p, q, q)$ (fuzzy left one-sided hypothesis), (1.9)
2. $\tilde{H}^{OR} = (p, p, q)$ (fuzzy right one-sided hypothesis), (1.10)
3. $\tilde{H}^{T} = (p, q, r)$ (fuzzy two-sided hypothesis), (1.11)

where $p < q < r \in \mathbb{R}$.

We assume that fuzziness is contained in a crisp random sample $X_1, \ldots, X_n$ having the probability distribution P_θ. For instance, to illustrate that, suppose we ask $n = 5$ pedestrians to guess approximately the measure of a given monument in the main square. Suppose we get the following answers: "about 10 m", "around

[1] In several researches in fuzzy methods, the triangle shape has been chosen by default to model fuzzy numbers, principally because of the shape's simplicity in terms of computations. For instance, Parchami et al. [7], and Filzmoser and Viertl [5] and others, used triangles in the context of fuzzy inference tests.

12 m", "about 9 or 10 m", "between 9 and 11 m", and "about 11 m". One can directly recognize the uncertainty in these answers. We model this fuzziness by fuzzy numbers as seen in Definition 1.2.

We obtain the following fuzzy random sample $\tilde{X} = (\tilde{X}_1, \ldots, \tilde{X}_n)$ with its corresponding membership function $\mu_{\tilde{X}}$ such that $\mu_{\tilde{X}}: \mathbb{R}^n \to [0, 1]^n$. For the latter, there exists a value $\mathbf{x}$ seen as an n-dimensional vector where it reaches 1. Then, we consider that its α-cuts are a closed compact and convex subset of $\mathbb{R}^n$.

Furthermore, we denote by ϕ a real valued function, $\phi: \mathbb{R}^n \to \mathbb{R}$. Let $\tilde{Z}$ be a fuzzy number resulting from applying the function ϕ on the fuzzy random sample $\tilde{X}$, i.e., $\tilde{Z} = \phi(\tilde{X}_1, \ldots, \tilde{X}_n)$. Then, the membership function $\mu_{\tilde{Z}}$ of $\tilde{Z}$ is as follows:

$$\mu_{\tilde{Z}}(z) = \begin{cases} \sup\{\mu_{\tilde{X}}(\mathbf{x}) : \phi(\mathbf{x}) = z\} & \text{if} \quad \exists \mathbf{x} : \phi(\mathbf{x}) = z, \\ 0 & \text{if} \quad \nexists \mathbf{x} : \phi(\mathbf{x}) = z, \end{cases} \tag{1.12}$$

for all z $\in \mathbb{R}$. The α-cuts of $\tilde{Z}$ can be written as:

$$\tilde{Z}_\alpha = [\min_{\mathbf{x} \in \tilde{X}_\alpha} \phi(\mathbf{x}), \max_{\mathbf{x} \in \tilde{X}_\alpha} \phi(\mathbf{x})], \tag{1.13}$$

for all $\alpha \in (0, 1]$ [8].

1.4.1 The Fuzzy Tests by Confidence Intervals

A way of testing hypotheses is by using confidence intervals. To accomplish this task, we have to construct a two-sided confidence interval (in the case of a two-sided test) denoted by $\tilde{\Pi}$, as described in Berkachy and Donzé [4]. This interval is built on the considered fuzzy sample at the significance level δ. Let us start by defining the fuzzy confidence interval.

Definition 1.5 (Fuzzy Confidence Interval) Consider a random sample $X_1, \ldots, X_n$ of size n. Let $[\pi_1, \pi_2]$ be the symmetrical confidence interval for θ at the significance level δ. A fuzzy confidence interval $\tilde{\Pi}$ is a convex and normalized fuzzy set given by its α-cuts, $\tilde{\Pi}_\alpha = [\tilde{\Pi}_\alpha^L, \tilde{\Pi}_\alpha^R]$. Its left and right α-cuts are given respectively as follows:

$$\tilde{\Pi}_\alpha^L = \inf\{a \in \mathbb{R} : \exists x_i \in (\tilde{X}_i)_\alpha, \forall i = 1, \ldots, n, \text{such that} \quad \pi_1(x_1, \ldots, x_n) \leq a\},$$

$$\tilde{\Pi}_\alpha^R = \sup\{a \in \mathbb{R} : \exists x_i \in (\tilde{X}_i)_\alpha, \forall i = 1, \ldots, n, \text{such that} \quad \pi_2(x_1, \ldots, x_n) \geq a\}.$$

Its membership function $\mu_{\tilde{\Pi}}(x)$ is defined as

$$\mu_{\tilde{\Pi}}(x) = \sup\{\alpha I_{[\tilde{\Pi}_\alpha^L, \tilde{\Pi}_\alpha^R]} : \alpha \in [0, 1]\}.$$

The confidence interval described in Definition 1.5 is given for a two-sided test. The one related to the left one-sided test is written by $\tilde{\Pi}_\alpha = [\tilde{\Pi}^L_\alpha, \infty]$, and $\tilde{\Pi}_\alpha = [-\infty, \tilde{\Pi}^R_\alpha]$ is the confidence interval for the right one-sided test.

Let $\mathbb{F}(\mathbb{R})$ be the space of all fuzzy sets of $\mathbb{R}$ and $\mathbb{FN}(\mathbb{R})$ the space of all fuzzy numbers. $\mathbb{FN}(\mathbb{R})$ is a subset of $\mathbb{F}(\mathbb{R})$. Consider $\tilde{\phi}$ a fuzzy test statistic. This function $\tilde{\phi}$, $\tilde{\phi} : (\mathbb{FN}(\mathbb{R}))^n \mapsto \mathbb{F}([0,1])$ is given by its α-cuts as follows:

$$\tilde{\phi}_\alpha(\tilde{X}_1, \ldots, \tilde{X}_n) = \begin{cases} \{0\} & \text{if } \left(\tilde{\Pi}_\alpha \setminus (\neg\tilde{\Pi})_\alpha\right) \cap \tilde{H}_0 = \theta_0; \\ \{1\} & \text{if } \left((\neg\tilde{\Pi})_\alpha \setminus \tilde{\Pi}_\alpha\right) \cap \tilde{H}_0 = \theta_0; \\ ([\alpha_1, \alpha_2]) \cup ([\alpha_3, \alpha_4]) & \text{if } \left(\tilde{\Pi}_\alpha \cap (\neg\tilde{\Pi})_\alpha\right) \cap \tilde{H}_0 = [A_L, A_R] \cup [R_L, R_R]; \\ \emptyset & \text{if } \left(\tilde{\Pi}_\alpha \cup (\neg\tilde{\Pi})_\alpha\right) \cap \tilde{H}_0 = \emptyset; \end{cases} \tag{1.14}$$

where $\alpha_1, \alpha_2, \alpha_3, \alpha_4 \in [0,1]$, $\alpha_1 \le \alpha_2$, $\alpha_3 \le \alpha_4$ and $A_L = \inf(\tilde{H}_0 \cap \tilde{\Pi}_\alpha)$, $A_R = \sup(\tilde{H}_0 \cap \tilde{\Pi}_\alpha)$, $R_L = \inf(\tilde{H}_0 \cap \neg\tilde{\Pi}_\alpha)$ and $R_R = \sup(\tilde{H}_0 \cap \neg\tilde{\Pi}_\alpha)$. The values $\alpha_1, \alpha_2, \alpha_3$ and α_4 are the corresponding α-cuts at the points A_L, A_R, R_L and R_R. We note that "$\neg$" refers to the negation operation and the expression $\tilde{\Pi}_\alpha \setminus (\neg\tilde{\Pi})_\alpha$ means the relative complement of $(\neg\tilde{\Pi})_\alpha$ with respect to the set $\tilde{\Pi}_\alpha$.

Proposition 1 *Consider*

$$p_0 = \min(\mu_{\tilde{\Pi}}(A_L), \mu_{\tilde{\Pi}}(A_R)), \ q_0 = \mu_{\tilde{\Pi}}(\theta_0), \ r_0 = \max(\mu_{\tilde{\Pi}}(A_L), \mu_{\tilde{\Pi}}(A_R)),$$

$$p_1 = \min(\mu_{\neg\tilde{\Pi}}(R_L), \mu_{\neg\tilde{\Pi}}(R_R)), \ q_1 = \mu_{\neg\tilde{\Pi}}(\theta_0) \ \textit{and} \ r_1 = \max(\mu_{\neg\tilde{\Pi}}(R_L), \mu_{\neg\tilde{\Pi}}(R_R)).$$

We denote by $\tilde{D}_0$ the fuzzy set derived from the intersection of the fuzzy null hypothesis $\tilde{H}_0$ and the fuzzy confidence interval $\tilde{\Pi}_\alpha$ (the letter "D" refers to the decision and the number "0" refers to the null hypothesis H_0). Then, $\tilde{D}_0$ is a triangular fuzzy number given by $\tilde{D}_0 = (p_0, q_0, r_0)$.

In the same way, we denote by $\tilde{D}_1$ the fuzzy set derived from the intersection of $\tilde{H}_0$ and $\neg\tilde{\Pi}_\alpha$. Then, $\tilde{D}_1$ is a triangular fuzzy number given by $\tilde{D}_1 = (p_1, q_1, r_1)$.

A proof of this proposition is given in Berkachy and Donzé [4]. It implies that the membership function of the fuzzy test statistic $\tilde{\phi}$ can be written as a union of membership functions associated to decisions on the null and alternative hypotheses. This membership function is given as follows:

$$\mu_{\tilde{\phi}}(t) = \mu_{\tilde{D}_0}(t)\ I_{\{\text{don't reject } H_0\}}(t) + \mu_{\tilde{D}_1}(t)\ I_{\{\text{don't reject } H_1\}}(t), \ t \in \{0, 1\}. \tag{1.15}$$

Fuzzy Decision Rule Let supp $\tilde{\Pi} = \{x \in \mathbb{R} : \mu_{\tilde{\Pi}}(x) > 0)\}$ be the support of the fuzzy confidence interval $\tilde{\Pi}$ and core $\tilde{\Pi} = \{x \in \mathbb{R} : \mu_{\tilde{\Pi}}(x) = 1)\}$ its kernel. Using all the above information, the fuzzy decision rule on θ is written as follows:

$$\tilde{\phi}(\tilde{X}_1, \ldots, \tilde{X}_n) = \begin{cases} \{1|H_0, \quad 0|H_1\} & \text{if } \theta_0 \in \text{core } \tilde{\Pi}; \\ \{0|H_0, \quad 1|H_1\} & \text{if } \theta_0 \notin \text{supp } \tilde{\Pi}; \\ \{\mu_{\tilde{D}_0}(t)|H_0, \mu_{\tilde{D}_1}(t)|H_1\} & \text{if else}; \end{cases} \tag{1.16}$$

where $|H_0$ and $|H_1$ indicate that the decision is related to the null and the alternative hypotheses, respectively, while 0 and 1 indicate "rejection" and "acceptance". In the third case, the decision rule is seen as a degree of conviction of accepting H_0 with $\mu_{\tilde{D}_0}(t)$ or rejecting it with $\mu_{\tilde{D}_1}(t)$.

1.4.2 The Fuzzy p-Value

According to the propositions and proofs given in Berkachy and Donzé [3], we provide the fuzzy p-value related to the case where both data and hypotheses are considered fuzzy. Its α-cuts are shown, along with the decision rule associated with (1.3), (1.4), and (1.5). The α-cuts of a fuzzy p-value can be calculated by the following proposition:

Proposition 2 *Given a test procedure based on the fuzziness of data and hypotheses and considering the three rejection regions* (1.6), (1.7), *and* (1.8), *the α-cuts of the fuzzy p-value $\tilde{p}$ are given by:*

1. $\tilde{p}_\alpha = [P_{\theta_R}(T \leq \tilde{t}^L_\alpha), P_{\theta_L}(T \leq \tilde{t}^R_\alpha)];$ (1.17)
2. $\tilde{p}_\alpha = [P_{\theta_L}(T \geq \tilde{t}^R_\alpha), P_{\theta_R}(T \geq \tilde{t}^L_\alpha)];$ (1.18)
3. $$\tilde{p}_\alpha = \begin{cases} [2P_{\theta_R}(T \leq \tilde{t}^L_\alpha), 2P_{\theta_L}(T \leq \tilde{t}^R_\alpha)] & \text{if} \quad A_l > A_r, \\ [2P_{\theta_L}(T \geq \tilde{t}^R_\alpha), 2P_{\theta_R}(T \geq \tilde{t}^L_\alpha)] & \text{if} \quad A_l \leq A_r; \end{cases} \tag{1.19}$$

for all $\alpha \in (0, 1]$, where $\tilde{t}^L_\alpha$ and $\tilde{t}^R_\alpha$ are the left and right α-cuts of $\tilde{t} = \phi(\tilde{X}_1, \ldots, \tilde{X}_n)$; θ_L and θ_R are the α-cuts of the boundary of $\tilde{H}_0$; P_θ is the probability distribution of T given θ; A_l is the area under the membership function $\mu_{\tilde{t}}$ of the fuzzy number $\tilde{t}$ on the left side of the median of the distribution of the test statistic T; and A_r is the one on the right side. In this case, one has to decide on which side the median is located based on the largest amount of fuzziness.

Proof 1 The detailed proof is given in Berkachy and Donzé [2].

Decision Rule According to the three-decision problem by the Neyman–Pearson statement, the decision rule for a given test at the significance level δ, is given as follows:

- if $\tilde{p}_{\alpha}^{R} < \delta$, the null hypothesis is rejected;
- if $\tilde{p}_{\alpha}^{L} > \delta$, the null hypothesis is not rejected;
- if $\delta \in [\tilde{p}_{\alpha}^{L}, \tilde{p}_{\alpha}^{R}]$, both null and alternative hypothesis are neither rejected nor not rejected.

1.5 Defuzzification of the Fuzzy Decisions by the Signed Distance

The signed distance operator was first defended by Yao and Wu [9]. Berkachy and Donzé [1] used it extensively in evaluating linguistic questionnaires. We open the section with the definition of the signed distance measure. This will be seen as a nice tool in the process of defuzzification of the fuzzy decision obtained in Sect. 1.4.1 from one side, and the defuzzification of the fuzzy p-value described in Sect. 1.4.2 from another one.

1.5.1 The Signed Distance

Let us first define the signed distance measure between two fuzzy sets.

Definition 1.6 The signed distance measured of a real value $a \in \mathbb{R}$ from the origin 0, $d_0(a, 0)$ is a, i.e., $d_0(a, 0) = a$.

We note that if $a < 0$, $-d_0(a, 0) = -a$. The signed distance between two real values a and $b \in \mathbb{R}$ is $d(a, b) = a - b$.

The signed distance between two fuzzy sets is:

Definition 1.7 The signed distance between $\tilde{D}$ and $\tilde{E}$ is

$$d(\tilde{D}, \tilde{E}) = \frac{1}{2} \int_0^1 [\tilde{D}_{\alpha}^{L} + \tilde{D}_{\alpha}^{R} - \tilde{E}_{\alpha}^{L} - \tilde{E}_{\alpha}^{R}] d\alpha.$$

This definition implies that the signed distance of a given fuzzy set measured from the fuzzy origin $\tilde{0}$ can be evaluated. We obtain the following:

Definition 1.8 The signed distance of $\tilde{D}$ measured from $\tilde{0}$ is:

$$d(\tilde{D}, \tilde{0}) = \frac{1}{2} \int_0^1 [\tilde{D}_{\alpha}^{L} + \tilde{D}_{\alpha}^{R}] d\alpha. \quad (1.20)$$

1.5.2 The Defuzzified Decision by Confidence Intervals

We reveal in this section the defuzzification of the fuzzy decision obtained according to the test related to confidence intervals seen in Sect. 1.4.1. Our purpose is to observe the testing problem as a linguistic variable. Therefore, we will consider the fuzzy test statistic decomposed into 2 linguistic terms $\tilde{L}_1$, $\tilde{L}_2$, i.e., rejection or acceptance of a given hypothesis, with their corresponding distances $d(\tilde{L}_1, \tilde{0})$ and $d(\tilde{L}_2, \tilde{0})$. Let δ_q be the indicator of an answer at the linguistic term L_q:

$$\delta_q = \begin{cases} 1 & \text{if the linguistic } L_q \text{ is true;} \\ 0 & \text{otherwise.} \end{cases} \tag{1.21}$$

We would like to express the signed distance measure O^{SGD} of the fuzzy decision $\tilde{D}$ as follows:

$$O^{SGD}(\tilde{D}) = \sum_{q=1}^{2} \delta_q d(\tilde{L_q}, \tilde{0}), \tag{1.22}$$

where $d(\tilde{L_q}, \tilde{0})$ is the signed distance of $\tilde{L_q}$.

Furthermore, we mention that $\tilde{L_1}$ and $\tilde{L_2}$ are nothing but $\tilde{D}_0$ and $\tilde{D}_1$, respectively, with their corresponding signed distances $d(\tilde{L}_1, \tilde{0}) = d(\tilde{D}_0, \tilde{0})$ and $d(\tilde{L}_2, \tilde{0}) = d(\tilde{D}_1, \tilde{0})$. We note that the signed distance of a triangular fuzzy number $\tilde{X} = (a, b, c)$ is given by $d(\tilde{X}, \tilde{0}) = \frac{1}{4}(a + 2b + c)$, and using this property, we get:

$$d(\tilde{L}_1, \tilde{0}) = \frac{1}{4}((\mu_{\tilde{\Pi}}(A_L) + 2\mu_{\tilde{\Pi}}(\theta_0) + \mu_{\tilde{\Pi}}(A_R))); \tag{1.23}$$

$$d(\tilde{L}_2, \tilde{0}) = \frac{1}{4}((\mu_{\neg\tilde{\Pi}}(R_L) + 2\mu_{\neg\tilde{\Pi}}(\theta_0) + \mu_{\neg\tilde{\Pi}}(R_R))). \tag{1.24}$$

Using Eqs. (1.16), (1.21), and (1.22), we obtain the crisp decision of the fuzzy hypothesis test. This decision can be written as follows:

$$\left(O^{SGD} \circ \tilde{\phi}\right)(\tilde{D}) = \begin{cases} 1 & \text{if } \mu_{\tilde{\Pi}}(\theta_0) = 1 \text{ (No rejection of } H_0\text{, rejection of } H_1\text{);} \\ 0 & \text{if } \mu_{\tilde{\Pi}}(\theta_0) = 0 \text{ (Rejection of } H_0\text{, no rejection of } H_1\text{);} \\ \delta_1 \cdot d(\tilde{D}_0, \tilde{0}) + \delta_2 \cdot d(\tilde{D}_1, \tilde{0}) & \text{if else.} \end{cases} \tag{1.25}$$

We note that in the case of crisp hypotheses, our method can be perfectly adopted as well. Further information is given in Berkachy and Donzé [4].

Decision Rule At the confidence level $1 - \delta$, the decision rule given in this case is as follows:

- we reject the null hypothesis if $d(\tilde{D}_0, \tilde{0}) < d(\tilde{D}_1, \tilde{0})$;
- we do not reject the null hypothesis if $d(\tilde{D}_0, \tilde{0}) > d(\tilde{D}_1, \tilde{0})$;
- both null and alternative hypotheses are neither rejected nor not rejected (a rare case) if $d(\tilde{D}_0, \tilde{0}) = d(\tilde{D}_1, \tilde{0})$. In this case, we will not be able to decide whether or not to reject the null hypothesis.

1.5.3 The Defuzzified p-Value

The signed distance will be used now in defuzzifying the fuzzy p-value given by its α-cuts, as shown in Eqs. (1.17), (1.18) and (1.19) of Sect. 1.4.2. By applying Eq. (1.20), we obtain the following defuzzified p-values:

$$1.\ d(\tilde{p}, \tilde{0}) = \frac{1}{2} \int_0^1 (P_{\theta_R}(T \le \tilde{t}_\alpha^L) + P_{\theta_L}(T \le \tilde{t}_\alpha^R))\mathrm{d}\alpha; \tag{1.26}$$

$$2.\ d(\tilde{p}, \tilde{0}) = \frac{1}{2} \int_0^1 (P_{\theta_L}(T \ge \tilde{t}_\alpha^R) + P_{\theta_R}(T \ge \tilde{t}_\alpha^L))\mathrm{d}\alpha; \tag{1.27}$$

$$3.\ d(\tilde{p}, \tilde{0}) = \begin{cases} \frac{1}{2} \int_0^1 (2P_{\theta_R}(T \le \tilde{t}_\alpha^L) + 2P_{\theta_L}(T \le \tilde{t}_\alpha^R))\mathrm{d}\alpha, & \text{if} \quad A_l > A_r, \\ \frac{1}{2} \int_0^1 (2P_{\theta_L}(T \ge \tilde{t}_\alpha^R) + 2P_{\theta_R}(T \ge \tilde{t}_\alpha^L))\mathrm{d}\alpha, & \text{if} \quad A_l \le A_r. \end{cases} \tag{1.28}$$

Decision Rule The decision rule related to the defuzzified p-values is similar to the one of the classical approach. It is given as follows:

- if $d(\tilde{p}, \tilde{0}) < \delta$, the null hypothesis is rejected;
- if $d(\tilde{p}, \tilde{0}) > \delta$, the null hypothesis is not rejected with the degree of conviction $d(\tilde{p}, \tilde{0})$;
- if $d(\tilde{p}, \tilde{0}) = \delta$ (a rare case), one should decide whether or not to reject the null hypothesis.

When the fuzzy p-value and the significance level overlap i.e. $\delta \in \text{supp}(\tilde{p})$, making a decision becomes much more complicated, and a new rule should be added. Thus, at this point, one has to compare the degree of conviction $d(\tilde{p}, \tilde{0})$ of not rejecting the null hypothesis with the threshold 0.5, and we get:

- if $d(\tilde{p}, \tilde{0}) < 0.5$, we tend to reject the null hypothesis with a degree of conviction $1 - d(\tilde{p}, \tilde{0})$;
- if $d(\tilde{p}, \tilde{0}) > 0.5$, we tend to not reject the null hypothesis with a degree of conviction $d(\tilde{p}, \tilde{0})$;
- if $d(\tilde{p}, \tilde{0}) = 0.5$, one should decide whether or not to reject the null hypothesis.

1.6 Tests of the Mean: Numerical Examples with Real Data

We expose in this section the application of the classical and fuzzy approaches on real data. We present the setups and the tests.

1.6.1 The Setups

For the application on real data, we used an example of a survey called "Finanzplatz: Umfrage 2010" done by the Office of Economy of the Canton of Zurich. This survey intended to understand the present and expected state of firms in Zurich in 2010 from different points of view, such as business situation, gross profit, and employment. We note that the survey is composed of 234 observations, i.e., firms with their corresponding sampling weights, answering 21 questions. We display in Table 1.1 a screenshot of a bloc of questions taken from the survey. These questions are mostly categorical ones, each having 5 possible answers (1 bad to 5 excellent). We call these categories "linguistics", and analogously such surveys are called "linguistic questionnaires".

The objective of this section is to show the usability of the hypotheses testing approach described in Sects. 1.4 and 1.5. We would like to have information about the present state of firms and about the demand for their services or products. We will then perform some tests on the mean in order to obtain this information. We can easily see that the answers for such questions are exposed to fuzziness and uncertainty, and therefore, a convenient treatment such as the fuzzy testing approach should be chosen. We propose to model the categories by fuzzy numbers. For the sake of simplicity, we will use triangular fuzzy numbers only, but the approach with different shapes of fuzzy numbers can be similarly conceivable. Table 1.2 displays the fuzzy numbers used. The fuzzy approach will be later compared to the classical one.

Table 1.1 Bloc of questions taken from the survey treated—application in Sect. 1.6

2 **Business situation**

General assessment

2.1 The present state of business is

bad		satisfactory		good
1	2	3	4	5
☐	☐	☐	☐	☐

Expectations

2.2 The expected state of business in 12 months is

worst		same		better
1	2	3	4	5
☐	☐	☑	☐	☐

Table 1.2 The possible linguistic terms and their corresponding fuzzy numbers—application in Sect. 1.6

Linguistic	Modality	Triangular fuzzy number
X_1	Bad	$\tilde{X}_1 = (0.5, 1, 1.5)$
X_2	Between bad and fair	$\tilde{X}_2 = (1, 2, 2.5)$
X_3	Fair	$\tilde{X}_3 = (2, 3, 4)$
X_4	Between fair and excellent	$\tilde{X}_4 = (3.5, 4, 5)$
X_5	Excellent	$\tilde{X}_5 = (4.5, 5, 5)$

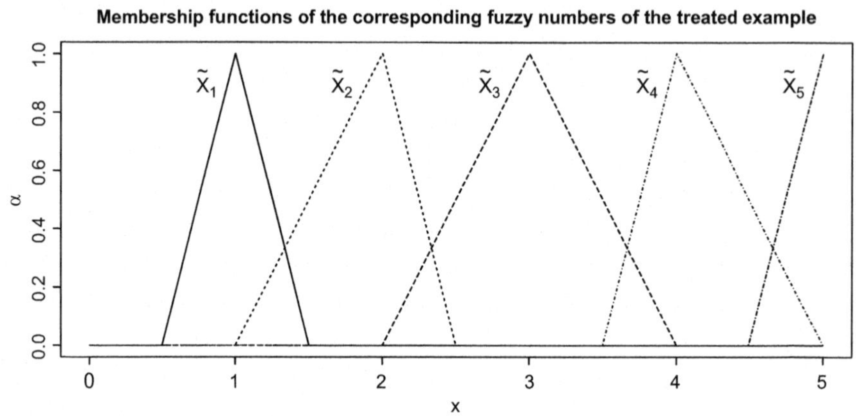

1.6.2 The Tests

(a) For the first test, we are interested in testing whether the average μ of the present state of business is approximately 3 (fair) or approximately smaller than 3 on the significance level δ=0.05. We write this left one-sided test as follows:

$$\tilde{H}_0 : \ \mu \text{ is approximately } \mu_0 = 3 \quad \text{vs.}$$
$$\tilde{H}_1 : \ \mu \text{ is approximately smaller than } \mu_0 = 3.$$

In our situation, and as we stated previously, the hypotheses as well as the data are considered fuzzy. The hypotheses are modelled by triangular fuzzy numbers and are given as:

$$\tilde{H}_0^T = (2.9, 3, 3.1) \quad \text{vs.} \quad \tilde{H}_1^{OL} = (3, 5, 5),$$

with their corresponding α-cuts:

$$(\tilde{H}_0^T)_\alpha = \begin{cases} (\tilde{H}_0^T)_\alpha^L = 2.9 + 0.1\alpha; \\ (\tilde{H}_0^T)_\alpha^R = 3.1 - 0.1\alpha. \end{cases} \tag{1.29}$$

$$(\tilde{H}_1^{OL})_\alpha = \begin{cases} (\tilde{H}_1^{OL})_\alpha^L = 3 + 2\alpha; \\ (\tilde{H}_1^{OL})_\alpha^R = 5. \end{cases} \tag{1.30}$$

First, we fuzzify the data and we get the fuzzy random sample $\tilde{X}_1, \ldots, \tilde{X}_{n=234}$. We then calculate the corresponding fuzzy sample mean $\overline{\tilde{X}}$. We get the tuple $\overline{\tilde{X}} = (3.056, 3.765, 4.442)$ with its α-cuts given by:

$$(\overline{\tilde{X}})_\alpha = \begin{cases} (\overline{\tilde{X}})_\alpha^L = 3.056 + 0.709\alpha; \\ (\overline{\tilde{X}})_\alpha^R = 4.442 - 0.677\alpha. \end{cases} \tag{1.31}$$

For the approach by confidence intervals, the next step is to construct the fuzzy confidence interval $\tilde{\Pi}$, as seen in Definition 1.5. To simplify our case, we assume that our fuzzified data set is derived from the normal distribution, and thus, the α-cuts of a fuzzy two-sided confidence interval for the mean are as follows:

$$\tilde{\Pi}_\alpha = [\tilde{\Pi}_\alpha^L, \tilde{\Pi}_\alpha^R] = [(\overline{\tilde{X}})_\alpha^L - u_{1-\frac{\delta}{2}} \frac{\sigma}{\sqrt{n}}, (\overline{\tilde{X}})_\alpha^R + u_{1-\frac{\delta}{2}} \frac{\sigma}{\sqrt{n}}], \tag{1.32}$$

where n is the sample size, σ is the standard deviation; $u_{1-\frac{\delta}{2}}$ is the $1 - \frac{\delta}{2}$ ordered quantile from the standard normal distribution; $(\overline{\tilde{X}})_\alpha^L$ and $(\overline{\tilde{X}})_\alpha^R$ are the left and right α-cuts of the fuzzy sample average $\overline{\tilde{X}}$, respectively.

We have to mention that other theoretical distributions can be similarly used according to characteristics of each data set. For our left one-sided case, using the α-cuts of the fuzzy sample mean $(\overline{\tilde{X}})_\alpha$ (Eq. (1.31)), the α-cuts of the fuzzy upper confidence interval for the mean on the confidence level $1 - 0.05$ are given as follows:

$$\tilde{\Pi}_\alpha = \left(-\infty, \tilde{\Pi}_\alpha^R\right] = \left(-\infty, (\overline{\tilde{X}})_\alpha^R + u_{1-\delta} \frac{\sigma}{\sqrt{n}}\right] = (-\infty, 4.546 - 0.677\alpha], \tag{1.33}$$

where $n = 234, \sigma = \sqrt{0.9445}, u_{1-\delta} = u_{0.95} = 1.64$.

We then study the intersection between the fuzzy confidence intervals and the fuzzy null hypothesis. Table 1.3 gives a graph of the case. The value of the membership functions at the intersection points are given as follows:

$$\mu_{\tilde{\Pi}}(A_L) = 1; \ \mu_{\tilde{\Pi}}(\mu_0) = 1; \ \mu_{\tilde{\Pi}}(A_R) = 1;$$

$$\mu_{\neg\tilde{\Pi}}(R_L) = 0; \ \mu_{\neg\tilde{\Pi}}(\mu_0) = 0; \ \mu_{\neg\tilde{\Pi}}(R_R) = 0.$$

Therefore, we can construct the fuzzy numbers $\tilde{D}_0$ and $\tilde{D}_1$, as described in Proposition 1, and we obtain $\tilde{D}_0 = (1, 1, 1)$ and $\tilde{D}_1 = (0, 0, 0)$ with their corresponding signed distances $d(\tilde{D}_0, \tilde{0}) = 1$ and $d(\tilde{D}_1, \tilde{0}) = 0$ related to "not

Table 1.3 The results of the crisp test and fuzzy one made using confidence intervals, and the membership function of the fuzzy p-value corresponding to the variable "The present state of business"—application in Sect. 1.6

The present state of business	The test results	The decisions
Crisp left one-sided test H_0: $\mu = 3$ vs. H_1: $\mu < 3$	• $T = 12.04 \quad > \quad t_l = -1.65142.$ • The p-value: 1.	H_0 is not rejected H_0 is totally not rejected
Fuzzy left one-sided test $\tilde{H}_0$: μ is approximately 3 vs. $\tilde{H}_1$: μ is smaller than 3	• The test by confidence intervals: $\tilde{D}_0 = (1, 1, 1), d(\tilde{D}_0, \tilde{0}) = 1,$ (Not reject H_0), $\tilde{D}_1 = (0, 0, 0), d(\tilde{D}_1, \tilde{0}) = 0,$ (Not reject H_1).	H_0 is not rejected
	• The defuzzified p-value: 0.967.	H_0 is not rejected

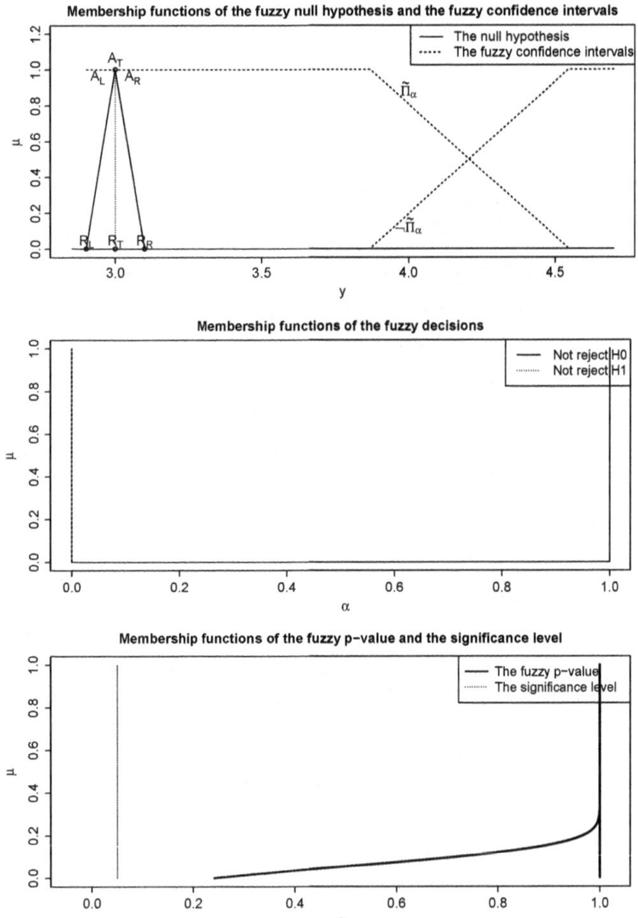

reject $\tilde{H}_0$" or "not reject $\tilde{H}_1$", respectively. Table 1.3 shows the obtained fuzzy decisions. We can now obtain the fuzzy test statistic, as given in Eq. (1.16). We can clearly see that $\mu_0 = 3 \in$ core $\tilde{\Pi}$, and thus, the fuzzy test statistic is written as follows:

$$\tilde{\phi}(\tilde{X}_1, \ldots, \tilde{X}_{234}) = \{1|H_0, \quad 0|H_1\}.$$

We are interested in defuzzifying the fuzzy test statistic by the signed distance. From Eq. (1.25), since $\mu_{\tilde{\Pi}}(\mu_0) = 1$, we totally decide to not reject the null hypothesis H_0, and to reject the alternative one H_1.

Decision According to the decision rule given in Sect. 1.5.2, we do not reject the null hypothesis H_0 since $d(\tilde{D}_0, \tilde{0}) = 1 > d(\tilde{D}_1, \tilde{0}) = 0$ at the 0.05 significance level.

Interpretation We do not reject the null hypothesis that the average of the present state of firms of our data set is approximately 3 (close to be fair). We reject the alternative hypothesis that this average is smaller than 3 (fair).

From another side, we would like to have the **fuzzy *p*-value** related to our case, and to defuzzify it, as described in Sect. 1.5.3. We mention that as a first task, one has to define the rejection region R. In our situation, we reject the null hypothesis H_0 if $T \leq t_l$, where t_l is such that $P(T \leq t_l) = \delta$ with t_l a quantile of the distribution of the test statistic T.

Then, we define the functions $\theta_1(\alpha)$ et $\theta_2(\alpha)$ as the limits of the integrals of the density function of the normal distribution. Using Eqs. (1.29) and (1.31), $\theta_1(\alpha)$ and $\theta_2(\alpha)$ are written as follows:

$$\theta_1(\alpha) = \frac{\overline{\tilde{X}}_\alpha^L - (\tilde{H}_0^T)_\alpha^R}{\sigma/\sqrt{n}} = \frac{-0.044 + 0.809 \times \alpha}{\sigma/\sqrt{n}} = -0.693 + 12.73 \times \alpha,$$

$$\theta_2(\alpha) = \frac{\overline{\tilde{X}}_\alpha^R - (\tilde{H}_0^T)_\alpha^L}{\sigma/\sqrt{n}} = \frac{1.542 - 0.777 \times \alpha}{\sigma/\sqrt{n}} = 24.27 - 12.23 \times \alpha.$$

We can now provide the fuzzy p-value associated to the normal density function given using its α-cuts $\tilde{p}_\alpha$ (see Eq. (1.17)) as follows:

$$\tilde{p}_\alpha = \left[\int_{-\infty}^{\theta_1(\alpha)} (2\pi)^{-\frac{1}{2}} \exp(\frac{-u^2}{2}) \mathrm{d}u, \int_{-\infty}^{\theta_2(\alpha)} (2\pi)^{-\frac{1}{2}} \exp(\frac{-u^2}{2}) \mathrm{d}u \right]. \tag{1.34}$$

We are now interested in defuzzifying this fuzzy p-value by the signed distance operator. According to Eq. (1.26), the defuzzified p-value is calculated by:

$$d(\tilde{p}, \tilde{0}) = \frac{1}{2} \int_0^1 (P_{\theta_R}(T \leq \tilde{t}_\alpha^L) + P_{\theta_L}(T \leq \tilde{t}_\alpha^R)) \mathrm{d}\alpha$$

$$= \frac{1}{2}\int_0^1 \Big(\int_{-\infty}^{-0.693+12.73\times\alpha} (2\pi)^{-\frac{1}{2}} \exp(\frac{-u^2}{2}) \mathrm{d}u \tag{1.35}$$

$$+ \int_{-\infty}^{26.49-14.45\times\alpha} (2\pi)^{-\frac{1}{2}} \exp(\frac{-u^2}{2}) \mathrm{d}u \Big) \mathrm{d}\alpha = 0.967.$$

Decision The defuzzified p-value previously calculated 0.967 is greater than the significance level $\delta = 0.05$. Consequently, we tend to strongly not reject the null hypothesis at this level with a degree of conviction of 0.967.

We note that the membership function of the fuzzy p-value $\tilde{p}$, as well as the results of the fuzzy tests are shown in Table 1.3. We add that we performed the same tests (by confidence intervals and the p-value) in the classical approach, and we obtained exactly the same decisions with a p-value of 1.

(b) For the second test, we would like to test whether the demand for the services or products of the firms compared to the last 12 months are approximately fair (3) or approximately more than fair (bigger than 3) on the significance level δ=0.05. The considered right one-sided test is:

$$\tilde{H}_0 : \ \mu \text{ is approximately } \mu_0 = 3 \quad \text{vs.}$$

$$\tilde{H}_1 : \ \mu \text{ is approximately bigger than } \mu_0 = 3.$$

As in the case of the first test, the hypotheses are considered fuzzy. In order to show different perspectives of the modelisation procedure, we model the hypotheses by other triangular fuzzy numbers given by the tuples $\tilde{H}_0^T = (2.6, 3, 3.2)$ and $\tilde{H}_1^{OR} = (3, 3, 5)$. Their α-cuts are as follows:

$$(\tilde{H}_0^T)_\alpha = \begin{cases} (\tilde{H}_0^T)_\alpha^L = 2.6 + 0.4\alpha; \\ (\tilde{H}_0^T)_\alpha^R = 3.2 - 0.2\alpha. \end{cases} \tag{1.36}$$

$$(\tilde{H}_1^{OR})_\alpha = \begin{cases} (\tilde{H}_1^{OR})_\alpha^L = 3; \\ (\tilde{H}_1^{OR})_\alpha^R = 5 - 2\alpha. \end{cases} \tag{1.37}$$

We calculate after the fuzzy sample mean $\overline{\tilde{Y}}$ of the fuzzy random sample $\tilde{Y}_1, \ldots, \tilde{Y}_{n=234}$, and we obtain $\overline{\tilde{Y}} = (2.598, 3.376, 4.194)$. Its α-cuts are given by:

$$(\overline{\tilde{Y}})_\alpha = \begin{cases} (\overline{\tilde{Y}})_\alpha^L = 2.598 + 0.778\alpha; \\ (\overline{\tilde{Y}})_\alpha^R = 4.194 - 0.818\alpha. \end{cases} \tag{1.38}$$

As for the first case, we will start the test by confidence intervals. This confidence intervals are calculated and we obtain the following α-cuts of the lower fuzzy confidence interval:

$$\tilde{\Pi}_\alpha = [\tilde{\Pi}^L_\alpha, +\infty) = [(\bar{\tilde{Y}})^L_\alpha - u_{1-\delta}\frac{\sigma}{\sqrt{n}}, +\infty] = [2.499 + 0.778\alpha, +\infty), \tag{1.39}$$

with $n = 234, \sigma = \sqrt{0.8365}, u_{1-\delta} = u_{0.95} = 1.64$.

We then construct the fuzzy intersection numbers, and we obtain the following triangular fuzzy numbers:

$$\tilde{D}_0 = (\mu_{\tilde{\Pi}}(A_L), \mu_{\tilde{\Pi}}(\mu_0), \mu_{\tilde{\Pi}}(A_R)) = (0.26, 0.64, 0.72),$$
$$\tilde{D}_1 = (\mu_{\neg\tilde{\Pi}}(R_R), \mu_{\neg\tilde{\Pi}}(\mu_0), \mu_{\neg\tilde{\Pi}}(R_L)) = (0.13, 0.36, 0.57),$$

with their respective signed distances, as seen in Eqs. (1.23) and (1.24):

$$d(\tilde{D}_0, \tilde{0}) = \frac{1}{4}(0.26 + 2 \times 0.64 + 0.72) = 0.567,$$
$$d(\tilde{D}_1, \tilde{0}) = \frac{1}{4}(0.13 + 2 \times 0.36 + 0.57) = 0.356.$$

We can now write the fuzzy test statistic as follows:

$$\tilde{\phi}(\tilde{Y}_1, \ldots, \tilde{Y}_{234}) = \{0.567|H_0, \quad 0.356|H_1\}.$$

Finally, we need a crisp decision as well. Since $\mu_{\tilde{\Pi}}(\mu_0) > 0$ and $\mu_{\tilde{\Pi}}(\mu_0) \neq 1$, we defuzzify the fuzzy decision by the signed distance, as presented in Eq. (1.25), and we obtain the following result:

$$\left(\mathcal{O}^{SGD} \circ \tilde{\phi}\right)(\tilde{Y}_1, \ldots, \tilde{Y}_{234}) = \delta_1 \cdot d(\tilde{D}_0, \tilde{0}) + \delta_2 \cdot d(\tilde{D}_1, \tilde{0}) = \delta_1 \cdot 0.567 + \delta_2 \cdot 0.356.$$

Decision By the decision rule of Sect. 1.5.2, since $d(\tilde{D}_0, \tilde{0}) = 0.567 > d(\tilde{D}_1, \tilde{0}) = 0.356$, we do not reject the null hypothesis H_0 with a degree of conviction of 0.567. We add that we do not reject the alternative one with a degree of 0.356.

Interpretation The signed distance 0.567 of not rejecting H_0 is larger than the signed distance 0.356 of rejecting it. The decision will then be to not reject the null hypothesis H_0 at the significance level 0.05. To sum up, we do not reject the null hypothesis that the average of the demand of services or products of the firms compared to the last 12 months is approximately fair (e.g. 3).

For the calculation of the **fuzzy *p*-value**, we know that we reject H_0 if $T \geq t_r$ with t_r a quantile of the distribution of the test statistic T is such that $P(T \geq t_r) = \delta$. Furthermore, with the bounds $\theta_1(\alpha)$ and $\theta_2(\alpha)$ of the normal density function integral given by $\theta_1(\alpha) = \frac{\overline{\tilde{Y}}_\alpha^R - (\tilde{H}_0^T)_\alpha^L}{\sigma/\sqrt{n}} = 26.66 - 20.37 \times \alpha$ and $\theta_2(\alpha) = \frac{\overline{\tilde{Y}}_\alpha^L - (\tilde{H}_0^T)_\alpha^R}{\sigma/\sqrt{n}} = -10.07 + 16.36 \times \alpha$, we obtain the following α-cuts of the fuzzy p-value $\tilde{p}$:

$$\tilde{p}_\alpha = \Big[\int_{\theta_1(\alpha)}^{\infty} (2\pi)^{-\frac{1}{2}} \exp(\frac{-u^2}{2}) du, \int_{\theta_2(\alpha)}^{\infty} (2\pi)^{-\frac{1}{2}} \exp(\frac{-u^2}{2}) du \Big], \tag{1.40}$$

with its corresponding defuzzified p-value, as described in Eq. (1.27) written as:

$$\begin{aligned} d(\tilde{p}, \tilde{0}) &= \frac{1}{2} \int_0^1 (P_{\theta_L}(T \geq \tilde{t}_\alpha^R) + P_{\theta_R}(T \geq \tilde{t}_\alpha^L)) d\alpha \\ &= \frac{1}{2} \int_0^1 \Big(\int_{\theta_1(\alpha)}^{\infty} (2\pi)^{-\frac{1}{2}} \exp\big(\frac{-u^2}{2}\big) du \int_{\theta_2(\alpha)}^{\infty} (2\pi)^{-\frac{1}{2}} \exp\big(\frac{-u^2}{2}\big) du \Big) d\alpha \\ &= 0.3077. \end{aligned}$$

Decision The defuzzified p-value 0.3077 is greater than the significance level $\delta = 0.05$. We deduce that we do not reject the null hypothesis H_0 with a degree of conviction of 0.3077. Since the fuzzy p-value and the significance level δ overlap, in other terms $0.3077 < 0.5$, the decision is difficult to take. Then, regarding the degree of conviction $1 - 0.3077 = 0.6923$, we can deduce that we reject the null hypothesis with a degree of conviction 0.6923.

Interpretation Therefore, we tend to reject the null hypothesis H_0 that the average of the demand of services or products of the firms compared to the last 12 months is approximately 3 at the significance level 0.05.

All the results related to this test, i.e., the crisp and fuzzy tests, the fuzzy decisions by the confidence intervals approach, and the fuzzy p-value, are given in Table 1.4.

1.7 Classical Approach vs. Fuzzy Approach

We intend in this section to compare the results obtained using the classical approach with the ones using the fuzzy approach. The ones related to the first case of Sect. 1.6 (see Table 1.3) show that the fuzzy and the classical crisp approaches gave exactly the same decisions regarding rejecting or not rejecting the null hypothesis. For instance, the null hypothesis H_0 is not rejected at the significance level 0.05 in both cases. We highlight that the interpretations of the decisions of both defuzzified and classical p-values were the same as well; the first one gave a defuzzified p-value of 0.967 and the second one of 1. At this point, it can be deduced that the fuzzy approach is more pessimistic than the classical one. Thus, this fuzzy p-value tends to "not reject" the null hypothesis at a lower degree than the classical one.

For the second test, the case is much more complicated since we can clearly see that the decisions obtained are in some points of view different between the classical

Table 1.4 The results of the crisp test and fuzzy one made using confidence intervals, and the membership function of the fuzzy p-value corresponding to the variable "The demand for our services/products compared to the last 12 months"—application in Sect. 1.6

The demand for our services/products compared to the last 12 months	The test results	The decisions
Crisp right one-sided test H_0: $\mu = 3$ vs. H_1: $\mu > 3$	• $T = 6.2898 > t_r = 1.65142$. • The p-value: $7.777e^{-10}$.	H_0 is rejected H_0 is rejected
Fuzzy right one-sided test $\tilde{H}_0$: μ is approximately 3 vs. $\tilde{H}_1$: μ is bigger than 3	• The test by confidence intervals: $\tilde{D}_0 = (0.26, 0.64, 0.72)$, $d(\tilde{D}_0, \tilde{0}) = 0.567$, (Not reject H_0), $\tilde{D}_1 = (0.13, 0.36, 0.57)$, $d(\tilde{D}_1, \tilde{0}) = 0.356$, (Not reject H_1). • The defuzzified p-value: 0.3077.	H_0 is not rejected We tend to reject H_0

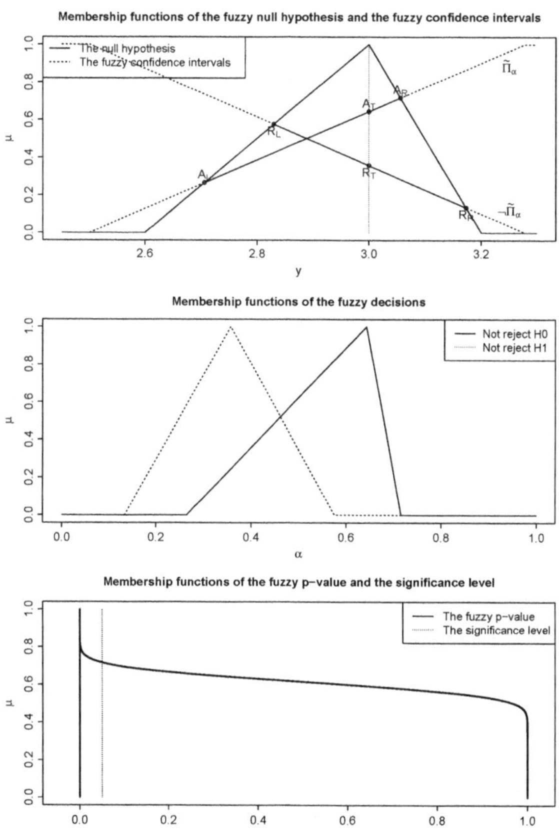

and the fuzzy approaches. The decision related to the classical approach is to reject the null hypothesis as well as the one related to the p-value ($7.777e^{-10}$). For the fuzzy approaches, the fuzzy decisions by the test using confidence intervals overlap, as seen in Figure 2 of Table 1.4. This leads to the idea of defuzzifying the decisions. The defuzzified decisions have a degree of conviction of 0.567 of not rejecting H_0 and a degree 0.356 of not rejecting H_1.

We mention that the concept of degree of conviction has been defined to further interpret the obtained fuzzy decisions. We add that even though we tend to not reject the null hypothesis according to the degree 0.567, we should say that this latter is not high enough (compared to 0.5, the evident threshold of comparison) to truly assume the rejection or no rejection of a given null hypothesis. This fact confirms the pessimistic side of the fuzzy approach regarding the classical one. Let us now look at the p-value. The third graph of Table 1.4 shows that the fuzzy p-value and the significance level overlap, and thus, a precise decision cannot be made. In such cases, the defuzzification of the fuzzy p-value can be of good use.

At the significance level $\delta = 0.05$, the degree of conviction of not rejecting the null hypothesis H_0 is 0.3077. Yet, since the $\tilde{p}$ and the significance level cross, we compare $d(\tilde{p}, \tilde{0})$ with 0.5 as seen in the decision rule of Sect. 1.5.3. While $0.3077 < 0.5$, we can obviously understand that we reject the null hypothesis with a degree $1-0.3077 = 0.6923$. Therefore, the decision is actually to reject the null hypothesis, as in the case of the classical approach.

We note that the degree of conviction is always assigned to the decision of "not rejecting" a given hypothesis. Consequently, our interpretations of the decisions should be always done according to this concept.

1.8 Recommendations of Use

Combining all the above information and discussions, we can clearly see that each of the classical and the fuzzy approaches have advantages and disadvantages. Thus, being able to provide guidelines for the use of each one of them is an absolute asset. Table 1.5 lists the pros and cons of both approaches.

Moreover, we have to highlight that obtaining a fuzzy decision is different than obtaining a crisp one from different perspectives. Even though the decisions that we make according to a crisp or a fuzzy test are in many cases the same, their interpretations are different. One of the big differences between both types of decisions is that in the classical approach, the decision is reduced to asserting whether or not we reject the hypothesis only. Contrariwise, for the fuzzy approach, a degree of conviction is defined for precisely interpreting the decisions. The applications of Sect. 1.6 and the discussion of Sect. 1.7 are the evidence of such statements. Thus, we can say that we can reject or not reject a given null hypothesis at a certain degree of conviction, and at the same time, we can reject or not reject the alternative one at a complementary degree of conviction.

Table 1.5 The advantages and disadvantages of the classical and the fuzzy approaches of hypotheses testing

	The classical approach	The fuzzy approach
Advantages	• Recommended for continuous variables. • Can be easily applied. • Can be used with many distributions.	• Suitable for discrete or continuous variables. • Suitable whenever fuzziness occurs. • Can be used with many distributions. • Seen as a generalization of the classical case. • Can be easily implemented using *R* functions.
Disadvantages	• Not suitable when fuzziness occurs.	• Difficult in terms of interpretation. • Concepts of fuzzy logic are mandatory.

At last, we mention that the classical approach is seen as an optimistic approach compared to the fuzzy one. Thus, this latter approach is obviously pessimistic. This conclusion is attributed to the fact that the classical approach tends to not reject a given hypothesis more strongly than the fuzzy approach. This optimism can be considered as an advantage in cases where the defended hypothesis is actually true. In other cases, this characteristic is seen as a disadvantage, especially since the decision would most probably be far from being realistic. For the pessimism of the fuzzy approaches, the discussion is exactly the opposite.

1.9 Conclusion

In this chapter, we reviewed fuzzy hypotheses testing approaches. We provided a test by confidence intervals as well by fuzzy p-values. In addition, since obtaining a crisp decision related to such situations is in many cases useful, we defuzzified the fuzzy decisions by a particular operator called the signed distance. We illustrated both approaches by an application on real data from a financial data set. This latter to show the business situation of firms of our sample. We tested our approaches on two variables describing the present state of businesses and the demand for their products or services compared to the last 12 months. We compared the results obtained from the fuzzy approach with the ones obtained from the classical approach.

We saw that in some cases both approaches gave the same decision regarding whether or not to reject a predefined hypothesis. In some other cases, one should be prudent about interpreting the decisions obtained from the fuzzy approaches particularly by carefully observing the corresponding degree of conviction. This latter is defined in order to obtain further precise information about our decisions.

For all these reasons, we believe that choosing between the classical and the fuzzy ones in testing hypotheses should be done prudently. Thus, in order to help the reader position between these approaches, we gave some guidelines for the use of each one. One additional interesting conclusion is that the application of our procedures is independent of the size and complexity of the data sets.

A future direction would be to investigate more statistical methods from the classical approach frequently used in economy and social sciences and where fuzziness eventually occurs, and extend them to the fuzzy environment.

References

1. Berkachy, R., & Donzé, L. (2016). Individual and global assessments with signed distance defuzzification, and characteristics of the output distributions based on an empirical analysis. In *Proceedings of the 8th International Joint Conference on Computational Intelligence - Volume 1: FCTA* (pp. 75–82).
2. Berkachy, R., & Donzé, L. (2017). Testing fuzzy hypotheses with fuzzy data and defuzzification of the fuzzy p-value by the signed distance method. In *Proceedings of the 9th International Joint Conference on Computational Intelligence (IJCCI 2017)* (pp. 255–264).
3. Berkachy, R., & Donzé, L. (2017). Defuzzification of the fuzzy p-value by the signed distance: Application on real data. In *Computational Intelligence. Studies in Computational Intelligence*. Berlin: Springer.
4. Berkachy, R., & Donzé, L. (2018). A new approach of testing fuzzy hypotheses by confidence intervals and defuzzification of the fuzzy decision by the signed distance. In *METRON*. Berlin: Springer.
5. Filzmoser, P., & Viertl, R. (2004). Testing hypotheses with fuzzy data: The fuzzy p-value. In *Metrika* (Vol. 59, pp. 21–29). Berlin: Springer.
6. Grzegorzewski, P. (2000). Testing statistical hypotheses with vague data. *Fuzzy Sets and Systems, 112*(3), 501–510.
7. Parchami, A., Taheri, S. M., & Mashinchi, M. (2010). Fuzzy p-value in testing fuzzy hypotheses with crisp data. *Statistical Papers, 51*(1), 209–226.
8. Viertl, R. (2011). *Statistical methods for fuzzy data*. Hoboken: Wiley.
9. Yao, J., & Wu, K. (2000). Ranking fuzzy numbers based on decomposition principle and signed distance. *Fuzzy Sets and Systems, 116*(2), 275–288.

Chapter 2
Interpolative Boolean Approach for Fuzzy Portfolio Selection

Aleksandar Rakićević, Pavle Milošević, Ana Poledica, Ivana Dragović, and Bratislav Petrović

2.1 Introduction

In the last few decades, fuzzy logic and other soft computing techniques have been gaining more attention in finance, especially in portfolio optimization [12, 30]. The aim of portfolio optimization is to select stocks and assign portfolio weights among selected stocks. Finding an optimal portfolio has been a challenging task for both investors and academics. Numerous portfolio optimization models have evolved over time (e.g. mean variance portfolio optimization models) in an attempt to maximize expected return and to minimize portfolio risk. To overcome the limitations of traditional models (e.g. portfolio size, practical constraints, investor's requirements, limited computation time, etc. [12]), many intelligent models based on soft computing and artificial intelligence techniques have been rapidly developing, particularly in the domain of portfolio selection.

Being uncertain and volatile in nature, financial markets are very difficult to model and predict. Focused only on statistical dependencies in the data, probabilistic approaches only partially capture the reality [8]. While traditional portfolio selection models are not entirely able to handle uncertainty, there are numerous studies showing the advantages of fuzzy theory when dealing with uncertainty, vagueness, and imprecision in financial markets [14]. Additionally, many research studies find that experts' knowledge and experience must be taken into consideration when selecting a portfolio, e.g. [24]. By using fuzzy approach, quantitative and qualitative analysis, experts' knowledge and investors' subjective opinions can be better integrated into a portfolio selection model [8]. Consequently, recent research

A. Rakićević (✉) · P. Milošević · A. Poledica · I. Dragović · B. Petrović
Faculty of Organizational Sciences, University of Belgrade, Belgrade, Serbia
e-mail: aleksandar.rakicevic@fon.bg.ac.rs

A. Meier et al. (eds.), *Applying Fuzzy Logic for the Digital Economy and Society*, Fuzzy Management Methods, https://doi.org/10.1007/978-3-030-03368-2_2

trends in the portfolio domain are to combine traditional mathematical modeling with fuzzy logic, genetic algorithms and other soft computing techniques [12, 30].

Interpolative Boolean algebra (IBA) is a [0,1]-valued generalization of Boolean algebra [32]. Unlike the classical fuzzy logic and other many-valued logics, IBA is consistent with Boolean frame, i.e. all Boolean laws, including the laws of contradiction and excluded middle are fulfilled in [0,1] interval. This has been accomplished by making the difference between the nature of attributes and their values. In fact, it requires the execution of predefined transformation rules before the values are introduced. IBA operates on whole [0,1] interval which coincides with the values of membership functions (i.e. the degree of fulfilment of some property). Consequently, IBA is considered to be a Boolean consistent tool adequate for solving problems immanent to fuzzy approach.

The aim of this chapter is to present the two decision-support techniques for portfolio selection based on IBA. These techniques are primarily based on logical aggregation [33], logical clustering [37] and IBA-based similarity/dissimilarity [26, 31]. First of all, we will discuss logical clustering (LC) approach for grouping companies according to the values of market valuation ratios [36]. LC approach is based on IBA dissimilarity measure and a standard hierarchical clustering algorithm. It is particularly suitable for comparison of multi-attribute objects, e.g. companies represented by intensities of their financial ratios. We will further reflect on IBA-based DuPont method for evaluation and deeper understanding of company's financial success. Traditional DuPont analysis decomposes the profitability ratio into several factors in order to explain profit drivers. Logical DuPont method [35] utilizes IBA framework to define and recognize the structure of distinctive factors (patterns) in order to construct a portfolio. The main benefit of this method is that it enables automated identification of a company's profit drivers interpreted in a fuzzy manner. In this research, we applied both logical clustering and logical DuPont on the S&P 500 stock market data to construct different stock portfolios and evaluate their performance.

The chapter is organized in follows: Sect. 2.2 is an overview of fuzzy methods applied for portfolio selection; the theoretical background of interpolative Boolean algebra is given in Sect. 2.3; Sect. 2.4 is devoted to IBA-based logical clustering method as well as experimental results and detailed analysis of the real case study, while Sect. 2.5 is dedicated to recently proposed logical DuPont method for the same case study. Section 2.6 concludes the chapter and provides the main guidelines for the future work.

2.2 Fuzzy Methods in Portfolio Selection

In this section, we first reflect on application of fuzzy theory in the existing traditional portfolio optimization models and its variants. Then, we highlight fuzzy based models for investment decision making introduced in recent papers.

One of the most frequently applied models for portfolio optimization is Markowitz mean-variance probabilistic model. The model assumes that investors are averse to risk and that returns of asset are random variable. The objective is to minimize the risk (defined by the variance) of the portfolio for the expected return (defined by the mean) of assets, i.e. to maximize the investment return with the given level of risk. Limitations of the mean-variance models are mostly reflected in the fact that security returns are often asymmetric, which makes the variance poor measure of risk. To overcome this limitation, Markowitz defined a semivariance model, some research employed skewness, while others directly used downside risk measure [17], etc.

Zadeh's theory of fuzzy sets has provided a new perspective on representation expected returns as fuzzy variables instead of random variables. Fuzzy mean-variance model was first proposed in [16] followed by fuzzy mean-semivariance model [17]. The Markowitz's model was also expanded into fuzzy possibilistic mean-variance model [41]. In the hybrid model [21], fuzzy set theory is applied to approximate expected value and variance of fuzzy returns. Beside the assumption that returns of securities can be represented as fuzzy random variables, the model proposed in [19] included investor's subjective opinions of the return rates. The multi-objective (return, risk and liquidity) optimization problem was also taken into consideration and modelled with random fuzzy returns to include investor's preference [20].

Traditional decision making methods have been further modified to support portfolio selection in fuzzy environment. The models such as fuzzy analytic hierarchy process (e.g. [28]), TOPSIS (e.g. [10]), etc. have been successfully applied. Fuzzy based models are being increasingly used for the development of expert decision support systems. Most research studies employ fuzzy systems based on fundamental and/or technical analysis for investment decision making. For the purpose of stock market trading, a Mamdani fuzzy logic system based on technical analysis was proposed [5]. Likewise, authors [3] used technical indicators as inputs when defining a complex TSK type fuzzy rule based system for stock price prediction. A system based on interval type-2 fuzzy logic with technical indicators also proved to be successful in predicting stock price movements [40]. Further, a fuzzy stock trading framework that takes into account both fundamental information and historical stock prices was evaluated in NASDAQ stock exchange [23]. For portfolio recommendation, authors [9] designed a fuzzy expert system on the basis of fundamental analysis ratios as well as qualitative criteria referring to experts' opinions on stocks listed at the Tehran Stock Exchange.

There are also many hybrid approaches that incorporate fuzzy and other soft computing techniques to address the problem of investment decision making. Some of them are neuro-fuzzy inference systems (e.g. [1]), genetic fuzzy systems (e.g. [15]), etc.

2.3 Theoretical Background: Interpolative Boolean Algebra

Interpolative Boolean algebra is a [0,1]-valued generalization of Boolean algebra (BA) [32]. Unlike the conventional fuzzy logic and other many-valued logics, IBA is consistent with Boolean frame, i.e. all Boolean laws, including the laws of contradiction and excluded middle, are satisfied in IBA. This has been accomplished by following the principle of structural functionality to distinguish between the nature of attributes and their values. Bearing in mind the structure of a logical expression and the nature of attributes, a different perspective on the evaluation of [0,1]-valued logical expressions is provided.

Formally, IBA consists of a symbolic and a value level. Any logical expression should be uniquely mapped to a generalized Boolean polynomial (GBP) on the symbolic level, while the expression is evaluated on the value level afterwards [33]. The Boolean consistency of IBA is an essential property from both mathematical and practical point of view. Therefore, IBA framework is used as a basis for numerous consistent fuzzy techniques and methods [25], as well as logical aggregation (LA) [33] and IBA similarity/dissimilarity measure [31, 36].

2.3.1 *Symbolic and Value Level of IBA*

On the symbolic level, IBA is identical to finite BA and can be represented as 4-tuple $\langle BA(\Omega), \wedge, \vee, \neg\rangle$, where $BA(\Omega)$ is a finite set of all logical expression formed using primary attributes $\Omega = \{a_1, \ldots, a_n\}$ and operators of conjunction, disjunction and negation. The main idea behind IBA symbolic level is to represent any element of $BA(\Omega)$ using atomic elements of IBA (atoms). Atomic elements are the simplest logical expression of $BA(\Omega)$, i.e. no element of $BA(\Omega)$ is incorporated in any atomic elements of IBA except itself and 0 constant. Atoms are mutually disjoint, while their sum is identically equal to 1. All other elements of $BA(\Omega)$ are built of two or more atomic attributes [32].

The presence/absence of particular atoms in a logical expression is described by the structure of the observed expression. The structure of an element of $BA(\Omega)$ is defined using structural vector, i.e. the n-dimensional binary vector where n represents the number of atoms. Therefore, any logical function on an IBA symbolic level is perceived as scalar product:

$$\varphi(a_1, \ldots, a_n) = A \cdot S(a_1, \ldots, a_n) \tag{2.1}$$

where A is the vector of atomic elements and $S(a_1, \ldots, a_n)$ is the structural vector of a logical expression.

Example 1 In case of BA generated by two attributes $\Omega = \{a_1, a_2\}$, IBA atomic elements are $a_1 \wedge a_2$, $a_1 \wedge \neg a_2$, $\neg a_1 \wedge a_2$ and $\neg a_1 \wedge \neg a_2$. For instance, the logical expression a_2 consists of atoms $a_1 \wedge a_2$ and $\neg a_1 \wedge a_2$, and its structural vectors is

$S_{a_1}(a_1, a_2) = [1\ \ 0\ \ 1\ \ 0]$. On the other hand, the corresponding structural vector for the logical expression $a_1 \vee a_2$ is $S_{a_1 \vee a_2}(a_1, a_2) = [1\ \ 1\ \ 1\ \ 0]$, since the expression contains 3 atoms: $a_1 \wedge a_2$, $a_1 \wedge \neg a_2$ and $\neg a_1 \wedge a_2$.

The structure of any complex element in $BA(\Omega)$ can be easily calculated directly on the basis of structures of its components. Therefore, the algebra in IBA framework is always Boolean.

In IBA framework, any logical expression is mapped to a generalized Boolean polynomial. GBP is a polynomial whose operators are standard plus and minus, and generalized product (GP), while variables are elements from the analyzed set of primary attributes. A GP can be any t-norm that satisfies the non-negativity condition [33], i.e. any function that has higher or equal value to Lukasiewicz t-norm and lower or equal to the minimum:

$$\max(0, a_1 + a_2 - 1) \leq a_1 \otimes a_2 \leq \min(a_1, a_2) \tag{2.2}$$

Although there are different realizations to be considered, the three cases of GPs are proven in practice. The minimum is used for aggregating attributes that are highly correlated and have similar meaning/nature. If the elements are of the similar meaning/nature but negatively correlated, Lukasiewicz t-norm is appropriate for GP. The product is used in case when attributes are independent.

Before introducing an operator suitable for GP, a set of predefined IBA transformation rules should be applied [33].

Example 2 In case of BA generated by three attributes $\Omega = \{a_1, a_2, a_3\}$, the logical function $(a_1 \wedge a_2) \vee (\neg a_1 \wedge a_3)$, may be transformed in GBP in the following manner:

$$\begin{aligned}
&((a_1 \wedge a_2) \vee (\neg a_1 \wedge a_3))^{\otimes} = \\
&\quad = (a_1 \wedge a_2)^{\otimes} + (\neg a_1 \wedge a_3)^{\otimes} - ((a_1 \wedge a_2) \wedge (\neg a_1 \wedge a_3))^{\otimes} \\
&\quad = a_1 \otimes a_2 + (1 - a_1) \otimes a_3 - a_1 \otimes a_2 \otimes (1 - a_1) \otimes a_3 \\
&\quad = a_1 \otimes a_2 + a_3 - a_1 \otimes a_3 - a_1 \otimes a_2 \otimes a_3 + a_1 \otimes a_2 \otimes a_1 \otimes a_3 \\
&\quad = a_1 \otimes a_2 + a_3 - a_1 \otimes a_3 - a_1 \otimes a_2 \otimes a_3 + a_1 \otimes a_2 \otimes a_3 \\
&\quad = a_1 \otimes a_2 + a_3 - a_1 \otimes a_3
\end{aligned} \tag{2.3}$$

In fact, the IBA transformation rules are utilized for the structure calculation. Any logical function may be modelled as a scalar product of the vector of atomic GBPs and structural vector. Hence, a logical expression to GBP can be transformed using either predefined IBA transformation rules or as a scalar product of the vector of atomic GBPs and a structural vector.

Finally, on IBA value level, attribute values are introduced and the appropriate GP operator is applied. Since all transformations are conducted in accordance with BA, all Boolean laws are preserved in the [0,1]-valued case.

2.3.2 Logical Aggregation

Since IBA allows Boolean consistent reasoning on the unit interval, it was employed as a basis for logical aggregation [33, 34]. LA is a consistent and transparent procedure for aggregating factors using logical expressions. It is conducted in two steps:

1. Normalization of attributes' values to [0,1] interval;
2. Aggregation of normalized values into resulting, globally representative value by means of a logical or a pseudo-logical function. A pseudo-logical function in IBA frameworks may be any linear convex combination of logical expressions expressed as GPBs.

Including logical dependencies between attributes in an aggregation process, LA provides a different perspective compared to traditional methods. One of the main benefits of LA is its interpretability, since LA functions are easy to understand and analyze. Secondly, LA supports non-monotone inference, which is not usual for aggregation procedures. Finally, LA is a very powerful aggregation tool since it generalizes many conventional aggregation operators such as weighted sum, arithmetic mean, OWA, etc. [33].

LA is utilized in many different application domains such as financial decision making [37], performance evaluation [18], etc.

2.3.3 IBA Dissimilarity Measure

Modelling similarity and dissimilarity using logic-based measures offer a different perspective in perceiving similarity. Logical relations, e.g. fuzzy implication, fuzzy bi-implication and IBA equivalence [31], are natural for measuring similarity/dissimilarity between attributes that describe the intensity of the properties. In IBA framework, a measure of dissimilarity is defined using the XOR relation between two objects [36] and the suitable GBP for d_{IBA} can be obtained using IBA transformation rules:

$$\begin{aligned} d_{IBA}(a,b) &= \left(a \underline{\vee} b\right)^{\otimes} = ((a \wedge \neg b) \vee (\neg a \wedge b))^{\otimes} \\ &= a + b - 2 \cdot a \otimes b \end{aligned} \tag{2.4}$$

Bearing in mind that only the same attributes of two objects can be compared, GP in IBA dissimilarity measure is implemented as the minimum. Therefore, the final expression for d_{IBA} is:

$$d_{IBA}(a,b) = a + b - 2 \cdot \min(a,b) \tag{2.5}$$

IBA XOR relation has a strong mathematical background and satisfies all the necessary conditions to be a measure of dissimilarity. It is a complement of IBA similarity measure s_{IBA} [31], i.e. $s_{IBA} = 1 - d_{IBA}$. More details about mathematical features and applications of d_{IBA} can be found in [36].

2.4 Logical Clustering for Portfolio Selection

The aim of this section is to introduce logical clustering method for portfolio selection problem. After a short overview on clustering techniques applied for portfolio selection, we thoroughly explain logical clustering and the steps for its application. Then, we introduce a particular stock selection problem, the portfolio construction algorithm and parameter settings. Finally, we evaluate resulting stock clusters and discuss portfolio performance.

Clustering is a valuable tool for stock market analysis since it discovering hidden patterns in financial data and extract significant information. Clustering has been applied to various problems in investment decision making.

Fuzzy logic is often applied together with clustering methods to assist in constructing optimal portfolios. In [13], a hybrid approach is proposed: a cluster analysis was employed for categorizing financial assets, the analytical hierarchy process for obtaining weights of the financial assets and fuzzy multi-objective linear programming model for portfolio selection. To cluster equity mutual funds, first Ward's method was used to compute the distance between clusters, and then k-means to minimize the variance within one cluster and to maximize the variance amongst other clusters [4]. The fuzzy optimization model was proposed to determine the optimal investment proportion of each cluster.

Another frequent use of clustering techniques for investment purposes is to evaluate company performance. In this application, companies are classified according to financial ratios obtained from company financial statements. As identified in [22], the most popular clustering methods applied with financial ratios are Ward's method and the k-means, which both use Euclidean distance. In [39], authors clustered the available financial ratios of different companies, and then a representative indicator of each cluster served as an evaluation criterion. In the clustering method, fuzzy equivalence relation is used to represent similarity between the financial ratios and to separate them into the clusters. Further, in [37], authors applied logical clustering approach using IBA-based exclusive disjunction to measure dissimilarity. To differentiate between undervalued and overvalued stocks, market valuation ratios were employed as evaluation criteria. Similarly, to enhance portfolio diversification, in [27], besides returns, authors made use of selected market valuation criteria. The well-known clustering methods k-means, self-organizing maps and fuzzy C-means were employed to categorize stocks, while selected stocks from different groups were used to build an efficient portfolio.

2.4.1 Logical Clustering

Logical clustering (LC) is a clustering technique based on a logic-based measure of proximity [36]. In LC algorithm, IBA relations are used within a standard hierarchical clustering algorithm. In fact, dissimilarity between objects is calculated using IBA dissimilarity measure and preferred LA operator. In this paper, the authors utilized the enhanced LC algorithm that allows automated decision making and meaningful cluster interpretation. The algorithm consists of the following steps:

1. Data normalization;
2. Calculating dissimilarities between objects using IBA relations;
3. Application of a linkage criterion;
4. Determining a number of clusters;
5. Automated interpretation.

Data Normalization Data normalization is a mathematical prerequisite for clustering using IBA-based relations. Since IBA is [0,1]-valued algebra, all attribute values must be mapped to the unit interval using the appropriate normalization function. It is particularly important that the chosen normalization function ensures that 'good' and 'bad' values of a certain attribute are well separated. The normalization function should also treat extreme values in a manner relevant to a problem. For instance, standard normalization functions scale data linearly within a unit interval (e.g. min-max normalization or scaling with the maximum value). Nonlinear functions based on data distribution are frequently applied. Finally, normalization functions may be modified or entirely provided by experts. In such a case, normalized values of attributes usually reflect the nature of the problem better, even though these functions may be biased.

Calculating Dissimilarities Between Objects Using IBA Relations Dissimilarity modelling in LC approach is based on an IBA dissimilarity measure and operator of LA. IBA dissimilarity measure treats all values from the unit interval equally and it is particularly useful for comparing attributes that represent an intensity of the properties. LA operator, however, makes possible to model different logical interactions among attributes and/or similarities. Formally, IBA dissimilarity modelling may be conducted as a simple attribute-by-attribute comparison along with suitable aggregation, or as a comparison on the level of the object [26].

Attribute-by-attribute comparison is perceived as a traditional manner of assessing diversity among multi-attribute objects. Dissimilarity between two objects $A = [a_1, \ldots, a_n]$ and $B = [b_1, \ldots, b_n]$ is modeled as a logical aggregation of IBA dissimilarities of individual attributes. Both traditional aggregation operators (e.g. mean, weighted sum, OWA) and logical relations can be used for aggregation of dissimilarities. This type of comparison has been successfully used in the LC algorithm in several papers [36, 37]. In these papers, the dissimilarities of individual attributes are aggregated using the following disjunction function:

$$d_{IBA}^{disj}(A, B) = \left(a_1 \underline{\vee} b_1\right) \vee \left(a_2 \underline{\vee} b_2\right) \vee \ldots \vee \left(a_n \underline{\vee} b_n\right) \qquad (2.6)$$

The previous function may be perceived as a strict one, since at least one significantly dissimilar attribute is sufficient to declare the two objects as diverse. Otherwise, IBA dissimilarities may be aggregated using conjunction that perceives two objects as diverse if none of their attributes are similar enough.

A comparison on the level of objects, however, is performed when it is important to model interactions and dependencies among attributes prior to dissimilarity measuring. This means that an object is uniquely represented using LA function before the application of IBA dissimilarity. This type of object comparison may not be conventional, but still meaningful and important in practice [26].

Application of a Linkage Criterion After the calculation of dissimilarities among objects, a conventional hierarchical clustering procedure is performed. Therefore, one of the common linkage criteria (e.g. single-linkage, average-linkage and complete-linkage) is applied to calculate dissimilarity between sets of objects.

Determining a Number of Clusters As in the classical hierarchical clustering, the number of clusters can be implicitly determined by using either statistical measures and methods (e.g. gap statistic, CH index), or expert judgement.

Automated Interpretation After establishing the number of clusters, each of them is represented by using its centroid. Centroid properties (i.e. values of its attributes) are further aggregated by using LA functions to identify cluster characteristics. These aggregation scores (i.e. cluster characteristics) are further used in the criterion function to obtain cluster interpretation as a basis for decision making.

2.4.2 Stock Selection Problem

Problem Setup and the Data For the purpose of this study, we have categorized 89 companies in meaningful clusters according to market valuation ratios. The aim is to differentiate between clusters in order to decide which one should be included in the stock portfolio, so that all the stocks from the selected clusters are part of a recommended portfolio.

In this study, the companies are selected based on the criteria of market capitalization and diverse industry sectors (e.g. Information Technology, Consumer Staples, Consumer Discretionary, Industrials, Materials, and Energy). All the companies are included in the S&P 500 index and traded on the American stock exchanges.

Table 2.1 Market valuation ratios

Ratio	Formula	Description
PE	$\frac{Market\ capitalization}{Net\ earnings}$	Represents a market value of one unit of the company's earnings (a dollar amount that an investor has to pay in order to buy one dollar of the company's earnings)
PB	$\frac{Market\ capitalization}{Shareholders'\ equity}$	Represents a market value of one unit of the shareholders' equity in a company (a dollar amount that an investor has to pay for one dollar of shareholders' equity)
PS	$\frac{Market\ capitalization}{Total\ revenue}$	Represents a market value of one unit of the company's revenue (a dollar amount that an investor has to pay for one dollar of the company's revenue)

The three common market valuation ratios (price per earnings (PE), price per book (PB) and price per sales (PS)) are used as the inputs for LC algorithm. These ratios are the indicators of market capitalization (market value of the company) relative to some financial statement items such as shareholders' equity, net earnings and total revenue. Subjectively, these ratios can indicate if a certain stock is overvalued (the ratio values are high) or undervalued (the ratio values are low). The formulas and basic information regarding ratios are given in Table 2.1.

The inputs to LC algorithm for portfolio selection are the financial reports data for the year 2015. It should be noted that companies from our dataset do not have the same reporting period, some of them ended their reporting period in March, June or September 2016. Clustering results and investment recommendations are evaluated monthly in a 6-month period after the financial report is released. In order to perform a fair evaluation, all daily closing prices are averaged on a monthly basis to avoid extreme values. It is important to emphasize that shorting (short selling) is allowed.

Portfolio Construction Algorithm In LC algorithm applied to the portfolio selection problem, each cluster is described by using two complementary LA functions C_i^+ and C_i^-. We then use the criterion function $cf^C = C_i^+ - C_i^-$ and the threshold parameter $tr^C \in [0, 1]$ to obtain recommended action. The portfolio construction algorithm presented in pseudo code is as follows:

Portfolio construction algorithm based on logical clustering

Step 1: $cf^C = C_i^+ - C_i^-, tr^C \in [0, 1]$
Step 2: IF $cf^C \geq tr^C$ THEN recommendedAction = BUY
Step 3: IF $cf^C \leq -tr^C$ THEN recommendedAction = SELL
Step 4: IF $-tr^C < cf^C < tr^C$ THEN recommendedAction = NO-ACTION

The low values of threshold parameter tr^C imply that the proposed portfolio is very diverse, while for high values of tr^C the portfolio consists of a smaller number of different stocks. Therefore, the value of parameter tr^C indicates the level of confidence which projects investors' risk aversion.

Parameter Settings Slightly modified min-max normalization function is used to scale the data on the unit interval. The maximum and the minimum values of each attribute are assigned after the removal of 5% of the extreme values from both ends of the scale. In other words, the four highest values of the observed attribute are fixed as 1, and the four lowest are fixed as 0. The rest of the values are mapped using the min-max normalization function.

The dissimilarities between the observed objects are assessed as a disjunction of attributes' IBA dissimilarities.

$$d_{IBA}^{disj}(A, B) = \left(PS_A \underline{\vee} PS_B\right) \vee \left(PB_A \underline{\vee} PB_B\right) \vee \left(PE_A \underline{\vee} PE_B\right) \tag{2.7}$$

This dissimilarity function evaluates two objects as diverse if they significantly differ in at least one attribute. According to IBA transformation rules, the logical expression is mapped to the following GBP:

$$\begin{aligned} d_{IBA}^{disj}(A, B) = {} & (PS_A + PS_B - 2 \cdot PS_A \otimes PS_B) + (PB_A + PB_B - 2 \cdot PB_A \otimes PB_B) \\ & + (PE_A + PE_B - 2 \cdot PE_A \otimes PE_B) + (PS_A + PS_B - 2 \cdot PS_A \otimes PS_B) \\ & \otimes (PB_A + PB_B - 2 \cdot PB_A \otimes PB_B) \otimes (PE_A + PE_B - 2 \cdot PE_A \otimes PE_B) \\ & - (PS_A + PS_B - 2 \cdot PS_A \otimes PS_B) \otimes (PB_A + PB_B - 2 \cdot PB_A \otimes PB_B) \\ & - (PS_A + PS_B - 2 \cdot PS_A \otimes PS_B) \otimes (PE_A + PE_B - 2 \cdot PE_A \otimes PE_B) \\ & - (PB_A + PB_B - 2 \cdot PB_A \otimes PB_B) \otimes (PE_A + PE_B - 2 \cdot PE_A \otimes PE_B) \end{aligned} \tag{2.8}$$

The transformation procedure is automatically executed using the software tool described in [25]. Bearing in mind the nature/meaning of the attributes, the GP for the dissimilarity measure should be executed in the following manner:

$$\begin{aligned} d_{IBA}^{disj}(A, B) = {} & (PS_A + PS_B - 2 \cdot \min(PS_A, PS_B)) + (PB_A + PB_B - 2 \cdot \min(PB_A, PB_B)) \\ & + (PE_A + PE_B - 2 \cdot \min(PE_A, PE_B)) + (PS_A + PS_B - 2 \cdot \min(PS_A, PS_B)) \\ & \cdot (PB_A + PB_B - 2 \cdot \min(PB_A, PB_B)) \cdot (PE_A + PE_B - 2 \cdot \min(PE_A, PE_B)) \\ & - (PS_A + PS_B - 2 \cdot \min(PS_A, PS_B)) \cdot (PB_A + PB_B - 2 \cdot \min(PB_A, PB_B)) \\ & - (PS_A + PS_B - 2 \cdot \min(PS_A, PS_B)) \cdot (PE_A + PE_B - 2 \cdot \min(PE_A, PE_B)) \\ & - (PB_A + PB_B - 2 \cdot \min(PB_A, PB_B)) \cdot (PE_A + PE_B - 2 \cdot \min(PE_A, PE_B)) \end{aligned} \tag{2.9}$$

The average linkage method is used as a linkage criteria as in the existing applications of LC [36, 37]. The number of clusters is determined by expert judgment, while centroids are interpreted using the two LA functions:

$$\begin{aligned} C_i^- &= PS_{C_i} \wedge \left(PB_{C_i} \vee PE_{C_i}\right) = \\ &= PS_{C_i} \otimes PB_{C_i} + PS_{C_i} \otimes PE_{C_i} - PS_{C_i} \otimes PB_{C_i} \otimes PE_{C_i} \\ &= PS_{C_i} \cdot PB_{C_i} + PS_{C_i} \cdot PE_{C_i} - PS_{C_i} \cdot PB_{C_i} \cdot PE_{C_i} \end{aligned} \tag{2.10}$$

$$\begin{aligned} C_i^+ &= \neg PS_{C_i} \wedge \left(\neg PB_{C_i} \vee \neg PE_{C_i}\right) = \\ &= 1 - PS_{C_i} - PB_{C_i} \otimes PE_{C_i} + PS_{C_i} \otimes PB_{C_i} \otimes PE_{C_i} \\ &= 1 - PS_{C_i} - PB_{C_i} \cdot PE_{C_i} + PS_{C_i} \cdot PB_{C_i} \cdot PE_{C_i} \end{aligned} \tag{2.11}$$

The higher values of the observed ratios indicate that a certain equity is highly valued at the market compared to others. Therefore, if the value of the function C_i^- is high, it is a sign that the cluster is comprised of overvalued stocks. In contrast, the clusters with high values of the C_i^+ function are identified as undervalued. In this study, we set the threshold value to $tr^C = 0.25$, thus ensuring the desired level of confidence.

2.4.3 Results and Discussion

LC algorithm separates stocks based on the three financial ratios in four clusters, as shown in Table 2.2. Cluster centroid properties (attributes), values of the criterion function and recommended actions are listed in the table. In Fig. 2.1 the clusters are easily identified.

The first cluster contains only one stock with extreme values for each attribute, i.e. PS and PB are equal to 1, while PE is 0. Consequently, this stock is clearly separated from all other clusters. However, it is suggested to sell this equity due to the exceptional value of PS. The stocks in the second cluster have high values of the ratios in general. Therefore, the recommended action is to sell all stocks from this cluster. The third cluster is the largest containing 65 stocks. Based on the criterion function, the investment recommendation is to purchase these equities. Finally, the last cluster contains 4 stocks with an exceptional average value of PE ratio. However, the average values of both PS and PB ratios are poor, so the criterion function does not give a strong signal to recommend a stock purchase at this point. Therefore, the resulting portfolio consists of 65 stocks bought and 20 stocks sold at month m. The portfolio is equally weighted. After the portfolio is constructed, we track its performance during the next 6 months. The highest achieved values of performance measures are highlighted in Table 2.3.

Table 2.2 Resulting clusters and investment recommendations

Cluster	Number of stocks	Average PS	Average PB	Average PE	cf^c	Recommended action
1	1	1.000	1.000	0.000	−1.000	SELL
2	19	0.728	0.675	0.563	−0.455	SELL
3	65	0.134	0.376	0.123	0.765	BUY
4	4	0.289	0.336	0.979	0.191	NO-ACTION

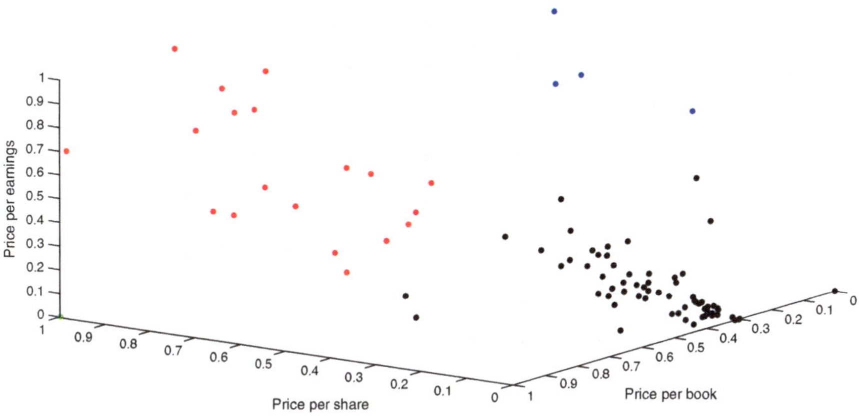

Fig. 2.1 Results of LC algorithm are presented using 3D scatter plot. The first cluster contains only one stock represented with green colour. The members of the second cluster are presented as red points. Black points represents stock from the third cluster. Blue colour is used to mark stocks that are elements of the fourth cluster

It is clear that the cumulative returns monotonically increase in both cases, while the hit rate is the maximal in month m+2. The monthly rate of return is dropping over time, indicating that the investment decision has the greatest effects shortly after it has been made. Regardless of the drop in monthly returns, the hit rate shows satisfactory performance staying above 60% during the whole time period. These results are in compliance with the well-known assumption of market efficiency that new information is incorporated into the stock price shortly after it is released.

It is interesting to notice that the portfolio achieves higher returns and hit rate when short selling is not allowed. Considering the average monthly returns of S&P 500 for year 2016 (0.87%) and for the 10-year period from 2006 to 2016 (0.49%), the proposed model shows satisfactory results. It is promising that the first month of portfolio performance (when the effects of investment decision are the greatest) significantly outperforms average S&P500 rates.

The results of this study provide us with some general recommendations for investment decision makers (e.g. portfolio managers, analysts, investors, etc.). When searching for new investment opportunities, an investor should focus on undervalued equities, i.e. the ones with lower values of PE, PS and PB ratios. It is interesting to note that PS has emerged as the most significant when constructing a criterion function for investment recommendations. This indicates that today's financial markets valuate companies more by market share than by their ability to generate earnings. Further, the results show that investors should avoid short selling. Finally, positive returns are highest shortly after investment decision has been made. This is in accordance with market efficiency hypothesis which states that all fundamental information is incorporated into price eventually. Therefore, the proposed method

Table 2.3 Portfolio performance of logical clustering method

	Number of stocks	Performance measure	*m*+1	*m*+2	*m*+3	*m*+4	*m*+5	*m*+6	Avg.
Portfolio	85	Cum. returns	1.84%	2.70%	3.15%	3.85%	4.52%	**4.69%**	/
(shorting		Mth. returns	**1.84%**	0.84%	0.44%	0.68%	0.65%	0.16%	0.77%
allowed)		Hit rate	61.18%	**69.41%**	67.06%	63.53%	62.35%	62.35%	64.31%
Portfolio	65	Cum. returns	2.75%	3.33%	3.55%	4.54%	**5.56%**	5.45%	/
(without		Mth. returns	**2.75%**	0.56%	0.21%	0.96%	0.98%	−0.1%	0.89%
shorting)		Hit rate	69.23%	**72.31%**	70.77%	66.15%	66.15%	64.62%	68.21%

Bold values represent maximum monthly values in observed 6-month period should only be used in the short term, and regularly updated with new fundamental information.

The proposed LC algorithm for stock selection is regarded as a general approach since the parameters can be adjusted according to an investor's preferences. The parameters tuning is possible regarding the time frame i.e. method can be applied to any time frame. The achieved results may also be improved using deeper financial analysis, especially for stocks within the larger clusters.

2.5 Logical DuPont Method

Logical DuPont is another IBA-based method for portfolio selection, which can be used as alternative to logical clustering explained previously. In this section, we explain original DuPont analysis as well as recently introduced logical DuPont method for the problem of portfolio selection. After data description and parameter settings, we discus results and give investment recommendations.

Financial analysis involves the selection, evaluation and interpretation of financial data and other pertinent information to assist in evaluating the operating performance and financial condition of a company [7]. Investors often use financial analysis as a tool to discover potential investment opportunities. DuPont analysis is a well-known method in financial analysis used to decompose a firm's return into its multiplicative components. This decomposition helps an investor to better understand the sources of return performances and enables him/her to differentiate between companies even when these performances are similar. Furthermore, there are some studies, like [38], that suggest DuPont components have explanatory power in predicting future profitability.

Return on equity (ROE) is a basic measure of a company's profitability, established as one of the most important criteria for investors to decide whether to invest in a particular company or not. Using DuPont decomposition, an investor can break down ROE into the product of two components: return on assets (ROA) and financial leverage (FL) ratios. Further, the ROA component can be disassembled into a product of net profit margin (NPM) and asset turnover (AT) ratio. The final disaggregation formula for ROE is:

$$ROE = ROA \cdot FL = NPM \cdot AT \cdot FL \tag{2.12}$$

Table 2.4 presents a detailed description of ratios and formulas for their calculation.

DuPont decomposition enables an investor to investigate the sources of ROE performance and to better understand the business model that generates company's earnings. The increase in overall ROE performance can be achieved by an increase in any of the three components. Even though the company can follow a certain strategy to boost its performance, it is often limited by industry constraints. Innovative industries, such as software companies and internet services are often

Table 2.4 DuPont components

Ratio	Formula	Description
ROE	$\frac{Net\ profit}{Shareholders'\ equity}$	Measures how effectively a company utilizes its equity to generate earnings.
ROA	$\frac{Net\ profit}{Total\ assets}$	Measures how effectively a company's assets are utilized to generate earnings.
NPM	$\frac{Net\ profit}{Revenue}$	Measures a company's level of effectiveness in generating profits from revenue.
AT	$\frac{Revenue}{Total\ assets}$	Measures the effectiveness of asset deployment in generating sales revenue.
FL	$\frac{Total\ assets}{Shareholders'\ equity}$	Measures how many of the company's assets are contributed by shareholders.

able to achieve high profit margins as the main source of their ROE performance. Well-established industries such as grocery stores, however, have limited ability to raise their profit margins and therefore tend to increase their turnover ratio or eventually to leverage their equity more aggressively. Similarly, some asset-intensive industries, such as oil companies, can hardly reach higher levels of asset turnover which limits their ability to increase the overall ROE. However, it is important to note that raising ROE performance by increasing ROA (through net profit margin or asset turnover) is more favourable than increasing financial leverage [11].

DuPont decomposition method was introduced to the field finance by Donaldson Brown back in 1918, and stayed for a long time in the shadow of the standard approach to financial ratio analysis. In the last few decades, the scientific community showed more interest in DuPont analysis, mainly in its practical application [6] and performance prediction [38]. As far as fuzzy and multi-valued logic application in DuPont analysis are concerned, to the best of our knowledge there are only two existing approaches. In [2], the author utilized fuzzy *C*-means (FCM) clustering algorithm to classify Turkish insurance companies based on DuPont model components as input variables. FCM and Celikyilmaz-Turksen's index are used to calculate membership values and identify the structural behaviour of DuPont based-MISO system. The other is a logic-based DuPont model developed for pattern identification in ROE's structure proposed in [35]. This approach is thoroughly explained in the following section.

2.5.1 Logical DuPont Method for Portfolio Selection

Logical DuPont decomposition method, introduced by Rakicevic [35], uses IBA theory to decompose ROE structure into basic structural elements—atoms. Each atom is a pattern that represents a business model, i.e. a way company achieved its earnings performance. The model enables investor to identify dominant atoms (patterns) and to recognize favourable behaviour among analysed companies. Further, an investor can use this information as a criterion for investment decision making.

The procedure for the logical DuPont method for portfolio construction is as follows:

1. Normalization of financial ratios;
2. Identification of ROE's structure using logical structural atoms;
3. Investment decisions making and a portfolio construction.

Normalization Normalization, fuzzification or some other kind of data transformation is a necessary prerequisite for any IBA-based technique. To normalize financial ratio data, we use min-max normalization, already explained in Sect. 2.4.1.

Identification of ROE's Structure This logical DuPont method uses three ROE components as primary BA attributes. Based on these attributes we create $2^3 = 8$ structural atoms using IBA theory. The atoms are in fact logical functions that represent business models (patterns in ROE's structure). These logical functions are further translated into corresponding GBPs, in order to be evaluated. Once an adequate *t*-norm is selected as a generalized product operator, the GBPs are easy to calculate. The obtained values represent the levels of fulfilment of each pattern in the ROE's structure. Patterns are presented in Table 2.5 along with corresponding logical functions, GBPs and interpretations.

Interpretation of the patterns is based on the conclusions reported in previous research studies. Soliman [38] claims that high NPM drives new entrants into the market or stimulates the existing rivals to imitate successful companies, which causes profit margins to revert to normal levels. He also states that, unlike profit margin, competition is less threatening to an efficient deployment of assets, because it is more difficult to imitate another's firm efficient production than another's firm new product idea. Soliman's work follows the works of [29]. All of them are consistent with the conclusion that AT is more persistent than NPM, and that

Table 2.5 Patterns in ROE's structure and their interpretations

Pattern	Logical model	GBP	Pattern interpretation
P1	NPM ∧ AT ∧ FL	NPM ⊗ AT ⊗ FL	Favourable pattern
P2	NPM ∧ AT ∧ ¬FL	NPM ⊗ AT ⊗ (1-FL)	Favourable pattern
P3	NPM ∧ ¬AT ∧ FL	NPM ⊗ (1-AT) ⊗ FL	Unfavourable pattern
P4	NPM ∧ ¬AT ∧ ¬FL	NPM ⊗ (1-AT) ⊗ (1-FL)	Gray pattern
P5	¬NPM ∧ AT ∧ FL	(1-NPM) ⊗ AT ⊗ FL	Favourable pattern
P6	¬NPM ∧ AT ∧ ¬FL	(1-NPM) ⊗ AT ⊗ (1-FL)	Favourable pattern
P7	¬NPM ∧ ¬AT ∧ FL	(1-NPM) ⊗ (1-AT) ⊗ FL	Unfavourable pattern
P8	¬NPM ∧ ¬AT ∧ ¬FL	(1-NPM) ⊗ (1-AT) ⊗ (1-FL)	Gray pattern

markets respond favourable to changes in AT which could be used as a predictor of future changes in firm's profitability.

Having all this in mind, we define our pattern interpretation logic as follows:

- All patterns that include high level of AT are considered as a favourable (patterns P1, P2, P5 and P6);
- All patterns that include low level of AT in combination with high level of FL are considered as unfavourable (patterns P3 and P7);
- All patterns that include low level of AT in combination with low level of FL are considered as gray—neither favourable nor unfavourable (patterns P4 and P8).

Investment Decision Making Once the ROE's structure is recognized, information about levels of pattern fulfilment is used to execute the following decision making algorithm for portfolio construction.

Portfolio construction algorithm based on logical DuPont method

Step 1:	favPatt = $[P1, P2, P5, P6]$, unfavPatt = $[P3, P7]$, neutPatt = $[P4, P8]$
Step 2:	$tr_1^{DP} \in [0, 1]$, $tr_2^{DP} \in [0, 1]$
Step 3:	IF SUM(favPatt) $\geq tr_1^{DP}$ THEN recommendedAction = BUY
Step 4:	IF SUM(unfavPatt) $\geq tr_1^{DP}$ THEN recommendedAction = SELL
Step 5:	IF SUM(grayPatt) $\geq tr_1^{DP}$
Step 5.1:	IF SUM(favPatt) $\geq tr_2^{DP}$ THEN recommendedAction = BUY
Step 5.2:	IF SUM(unfavPatt) $\geq tr_2^{DP}$ THEN recommendedAction = SELL
Step 6:	ELSE recommendedAction = NO-ACTION

In the algorithm, the idea is to add together the level of fulfilment for all patterns of the same type (favourable, unfavourable and gray) for each stock. To make an investment decision (to buy, to sell or to do nothing) the algorithm investigates whether there exists a dominant group of patterns. Further, we define primary and secondary dominance. A group of patterns is primarily dominant if their sum of levels of fulfilment is larger than the threshold parameter tr_1^{DP}. Likewise, a group is secondarily dominant if the corresponding sum is larger than tr_2^{DP}. In case when gray patterns are primarily dominant, algorithm investigates whether there is a secondarily dominant group of patterns. The higher levels of thresholds lead to more rigorous selection which is a characteristic of risk-averse investors.

2.5.2 Data and Parameter Settings

Logical DuPont method is applied to the same set of 89 companies used in the previous case study. The financial ratios are calculated from financial reports for the year 2015. Investment recommendations are evaluated using monthly averages of daily closing prices in a period of 6 months after the financial report is released publicly. It is important to underline that short selling is allowed as before.

The values of ROE components vary significantly in our dataset. This is a consequence of the fact that the companies operating in different industries often utilize significantly different business models. To separate attribute values in a meaningful manner, we apply min-max normalization function where min and max parameters are set by an expert.

Since NPM, AT and FL ratios represent distinctive aspects of a company's performance, we consider them as different in nature. Therefore, we propose the standard product as an adequate operator of GP.

As previously noticed, an investment decision making step of logical DuPont method procedure relies on two threshold parameters: tr_1^{DP} and tr_2^{DP}. Their values are set expert-wise on $tr_1^{DP} = 0.5$ which is equivalent to 50% of fulfilment and $tr_2^{DP} = 0.5$ as 25% of fulfilment.

2.5.3 *Results and Discussion*

On the basis of pattern recognition in ROE's structure, the portfolio construction algorithm (presented in Sect. 2.5.1) recommends 36 out of 89 companies for purchase and 8 for short selling. For the remaining 45 stocks it is suggested to take no action. In Table 2.6 we present the portfolio performance for a 6-month period. We also present the results when short selling is not allowed, i.e. DuPont method is used to select stocks only for purchase.

The portfolio performance results show that the applied portfolio construction algorithm was able to achieve a positive rate of both monthly and cumulative returns during the 6-month period. The cumulative rate of return is monotonically increasing, proving that the positive trend was captured, even though monthly rate of return decreases over time. As already explained in Sect. 2.4.3, the decrease in monthly rate is natural since the effects of investment decision should diminish over time. In case short selling is not allowed, the constructed portfolio shows exceptional performances with an average monthly rate of 2.47% for the first 2 months, and 1.08% for the whole 6-month period. It is important to note that both of these rates are greater than average monthly S&P 500 rate for the year 2016. These results are even more significant when compared to average monthly rate of S&P 500 index for 10-year period (0.42%).

Hit rate is constantly above 50% and in most of the cases the rate is above 65%. The average rate value of 67.04% indicates that logical DuPont is successful as a method for investment decision making. It is interesting that portfolio performances improve drastically when short selling is not allowed, which implies that the method better recognize perspective stocks unlike the stocks recommended for selling. We can conclude that DuPont method has the ability of deeper understanding of the relations in financial fundamentals that lead to positive trends in price returns.

Based on these results, it is possible to make some general investment recommendations. In analyzing ROE's structure, AT emerges as the most important ratio

Table 2.6 Portfolio performances of logical DuPont method

	Number of stocks	Performance measure	*m*+1	*m*+2	*m*+3	*m*+4	*m*+5	*m*+6	Avg.
Portfolio (shorting allowed)	44	Cum. returns	0.65%	1.86%	2.03%	2.23%	2.52%	**2.53%**	/
		Mth. returns	0.65%	**1.20%**	0.17%	0.20%	0.28%	0.01%	0.42%
		Hit rate	59.09%	70.45%	70.45%	68.18%	**72.73%**	61.36%	67.04%
Portfolio (without shorting)	36	Cum. returns	2.69%	5.00%	5.51%	5.90%	**6.67%**	6.60%	/
		Mth. returns	**2.69%**	2.25%	0.49%	0.37%	0.76%	−0.09%	1.08%
		Hit rate	66.67%	**77.78%**	**77.78%**	72.22%	**77.78%**	63.89%	72.67%

Bold values represent maximum monthly values in observed 6-month period for identifying favorable patterns. Therefore, investors should focus on companies with high levels of AT ratio, i.e. companies which generate revenues from the assets faster than others. As in the previous case study, it is suggested to avoid shorting as it decreases portfolio performance. As before, the results show high profitability in short-term period.

The presented portfolio construction algorithm is a general approach that is, the parameters can be fine-tuned with respect to investor's preferences. Also, the algorithm can be adjusted to work in different time frames.

2.6 Conclusion

In this chapter, we presented two portfolio selection methods: logical clustering and logical DuPont analysis. Both methods are based on interpolative Boolean algebra. IBA is a [0,1]-valued Boolean consistent framework that enables fuzzy realization of logical expressions. In order to prove their potential for stock selection problems, they were validated on real market data. For the purpose of this study, we used data on 89 companies included in S&P 500 from 6 different industries.

Logical clustering method employs IBA-based approach within a standard hierarchical clustering algorithm. Fuzzy theory, i.e. IBA as consistent fuzzy approach, enables finer differentiation between clusters. On the bases of market valuation ratios, LC can provide more meaningful clusters of stocks that should be included in a portfolio. The LC algorithm was further enhanced with LA functions for automated interpretation of each cluster. This advantage enables a deeper understanding of the cluster nature as well as meaningful investment recommendations, without the involvement of a human being. On the other hand, logical DuPont uses IBA theory to decompose ROE using structural atoms with the aim of discovering company's profit drivers. Consequently, this method also enables automated detection of ROE's structure and the identification of favourable behaviour patterns for investment recommendations.

The portfolios constructed with IBA-based methods achieved promising results. In general, theirs average monthly returns followed S&P 500 index performances. In cases when short selling was not allowed, the portfolios outperformed S&P 500. Moreover, both methods achieved high levels of hit rate, maintaining them above 60% during the whole test period. The overall results led us to the conclusion that logical clustering and logical DuPont decomposition methods are valuable tools that can be used for building automated systems for stock selection and portfolio construction.

It is also important to note that the chosen normalization function can be significant for the performance of described IBA-based methods. In this study, the company data from different industries are treated with the same normalization function. Further improvements could include specially adjusted normalization functions for each industry sector. In terms of logical DuPont analysis, the decomposition of ROE's structure includes a small number of structural atoms, so computation is not time consuming. However, in applications with large number of atoms computation time may be a possible limitation. The illustrated methods are general, which means the parameters can be adjusted according to the specific investor and his risk aversion.

Both logical clustering and logical DuPont have the same goal—to select stocks for a portfolio, even though they capture the financial ratios that are different in nature. In this research, we used the two methods separately. However, they can be complementary, e.g. selected clusters as parts of recommended portfolios can be the subject of deeper decomposition analysis in logical DuPont method.

References

1. Bagheri, A., Peyhani, H. M., & Akbari, M. (2014). Financial forecasting using ANFIS networks with quantum-behaved particle swarm optimization. *Expert Systems with Applications, 41*(14), 6235–6250.
2. Baskir, M. B. (2016). Type-1 fuzzy modeling for DuPont analysis on Turkish insurance sector. *Turkish Journal of Fuzzy Systems, 7*(1), 29–40.
3. Chang, P. C., & Liu, C. H. (2008). A TSK type fuzzy rule based system for stock price prediction. *Expert Systems with Applications, 34*(1), 135–144.
4. Chen, L. H., & Huang, L. (2009). Portfolio optimization of equity mutual funds with fuzzy return rates and risks. *Expert Systems with Applications, 36*(2), 3720–3727.
5. Chourmouziadis, K., & Chatzoglou, P. D. (2016). An intelligent short term stock trading fuzzy system for assisting investors in portfolio management. *Expert Systems with Applications, 43*, 298–311.
6. Dehning, B., & Stratopoulos, T. (2002). DuPont analysis of an IT-enabled competitive advantage. *International Journal of Accounting Information Systems, 3*(3), 165–176.
7. Drake, P. P., & Fabozzi, F. J. (2010). *The basics of finance: an introduction to financial markets, business finance, and portfolio management* (Vol. 192). New York: Wiley.
8. Fang, Y., Lai, K. K., & Wang, S. (2008). *Fuzzy portfolio optimization: Theory and methods* (Vol. 609). Berlin: Springer.
9. Fasanghari, M., & Montazer, G. A. (2010). Design and implementation of fuzzy expert system for Tehran Stock Exchange portfolio recommendation. *Expert Systems with Applications, 37*(9), 6138–6147.
10. Ferreira, L., Borenstein, D., Righi, M. B., & De Almeida Filho, A. T. (2018). A fuzzy hybrid integrated framework for portfolio optimization in private banking. *Expert Systems with Applications, 92*, 350–362.
11. Fridson M., & Alvarez, F. (2002). *Financial statement analysis: A practitioner's guide* (3rd ed.). New York: Wiley.
12. Ghosh, A., & Mahanti, A. (2014). Investment portfolio management: A review from 2009 to 2014. In *Proceedings of the 10th Global Business and Social Science Research Conference, Beijing* (pp. 23–24).
13. Gupta, P., Inuiguchi, M., & Mehlawat, M. K. (2011). A hybrid approach for constructing suitable and optimal portfolios. *Expert Systems with Applications, 38*(5), 5620–5632.
14. Gupta, P., Mehlawat, M. K., Inuiguchi, M., & Chandra, S. (2014). Fuzzy portfolio optimization. In *Studies in fuzziness and soft computing* (Vol. 316). Berlin: Springer.
15. Hadavandi, E., Shavandi, H., & Ghanbari, A. (2010). Integration of genetic fuzzy systems and artificial neural networks for stock price forecasting. *Knowledge-Based Systems, 23*(8), 800–808.

16. Huang, X. (2007). Portfolio selection with fuzzy returns. *Journal of Intelligent Fuzzy Systems, 18*(4), 383–390.
17. Huang, X. (2008). Mean-semivariance models for fuzzy portfolio selection. *Journal of Computational and Applied Mathematics, 217*(1), 1–8.
18. Latinovic, M., Dragovic, I., Bogojevic Arsic, V., & Petrovic, B. (2018) A fuzzy inference system for credit scoring using Boolean consistent fuzzy logic. *International Journal of Computational Intelligence Systems, 11*, 414–427.
19. Li, J., & Xu, J. (2009). A novel portfolio selection model in a hybrid uncertain environment. *Omega, 37*(2), 439–449.
20. Li, J., & Xu, J. (2013). Multi-objective portfolio selection model with fuzzy random returns and a compromise approach-based genetic algorithm. *Information Sciences, 220*, 507–521.
21. Li, X., Zhang, Y., Wong, H. S., & Qin, Z. (2009). A hybrid intelligent algorithm for portfolio selection problem with fuzzy returns. *Journal of Computational and Applied Mathematics, 233*(2), 264–278.
22. Linares-Mustaros, S., Coenders, G., & Vives-Mestres, M. (2017). Financial performance and distress profiles. From classification according to financial ratios to compositional classification. *Advances in Accounting, 40*, 1–10.
23. Lincy, G. R. M., & John, C. J. (2016). A multiple fuzzy inference systems framework for daily stock trading with application to NASDAQ stock exchange. *Expert Systems with Applications, 44*(C), 13–21.
24. Liu, Y. J., Zhang, W. G., & Xu, W. J. (2012). Fuzzy multi-period portfolio selection optimization models using multiple criteria. *Automatica, 48*(12), 3042–3053.
25. Milosevic, P., Petrovic, B., Radojevic, D., & Kovacevic, D. (2014). A software tool for uncertainty modeling using Interpolative Boolean algebra. *Knowledge-Based Systems, 62*, 1–10.
26. Milosevic, P., Poledica, A., Rakicevic, A., Dobric, V., Petrovic, B., & Radojevic, D. (2018). IBA-based framework for modeling similarity. *International Journal of Computational Intelligence Systems, 11*, 206–218.
27. Nanda, S. R., Mahanty, B., & Tiwari, M. K. (2010). Clustering Indian stock market data for portfolio management. *Expert Systems with Applications, 37*(12), 8793–8798.
28. Nguyen, T. T., & Gordon-Brown, L. (2012). Constrained fuzzy hierarchical analysis for portfolio selection under higher moments. *IEEE Transactions on Fuzzy Systems, 20*(4), 666–682.
29. Nissim, D., & Penman, S. (2001). Ratio analysis and equity valuation: From research to practice. *Review of Accounting Studies, 6*(1), 109–154.
30. Pareek, M. K., & Thakkar, P. (2015). Surveying stock market portfolio optimization techniques. In: *Proceedings of the 5th Nirma University International Conference on Engineering (NUiCONE 2015)* (pp. 1–5). Piscataway: IEEE.
31. Poledica, A., Milosevic, P., Dragovic, I., Petrovic, B., & Radojevic, D. (2015). Modeling consensus using logic-based similarity measures. *Soft Computing, 19*(11), 3209–3219.
32. Radojevic, D. (2000). New [0,1]-valued logic: A natural generalization of Boolean logic. *Yugoslav Journal of Operations Research, 10*(2), 185–216.
33. Radojevic, D. (2008). Logical aggregation based on interpolative Boolean algebra. *Mathware & Soft Computing, 15*, 125–141.
34. Radojevic, D., & Petrovic, B. (2001). Logical aggregation of [0,1]-valued logic. In R. S. Stankovic & C. Moraga (Eds.), *Proceedings of a Workshop on Computational Intelligence, Theory and Application* (pp. 69–80). Nis: Elektronski fakultet.
35. Rakicevic, A., Milosevic, P., Petrovic, B., & Radojevic, D. (2016). DuPont financial ratio analysis using logical aggregation. In V. E. Balas, L. C. Jain, & B. Kovacevic (Eds.) *Soft computing applications. Advances in intelligent systems and computing* (Vol. 357, pp. 727–739). Berlin: Springer.

36. Rakicevic, A., Nesic, I., & Radojevic, D. (2013). A novel approach to hierarchical clustering based on logical measure of dissimilarity. In N. Mladenovic, G. Savic, M. Kuzmanovic, D. Makajic-Nikolic, & M. Stanojevic (Eds.), *Proceedings of the 11th Balkan Conference on Operational Research* (pp. 147–155). Belgrade: Faculty of Organizational Sciences.
37. Rakićević, A., Simeunović, V., Petrović, B. & Milić, S. (2018). An automated system for stock market trading based on logical clustering. *Tehnički vjesnik, 25*(4), 970–978.
38. Soliman, M. T. (2008). The use of DuPont analysis by market participants. *The Accounting Review, 83*(3), 823–853.
39. Wang, Y. J., & Lee, H. S. (2008). A clustering method to identify representative financial ratios. *Information Sciences, 178*(4), 1087–1097.
40. Zarandi, M. F., Rezaee, B., Turksen, I. B., & Neshat, E. (2009). A type-2 fuzzy rule-based expert system model for stock price analysis. *Expert Systems with Applications, 36*(1), 139–154.
41. Zhang, W. G. (2007). Possibilistic mean-standard deviation models to portfolio selection for bounded assets. *Applied Mathematics and Computation, 189*(2), 1614–1623.

Chapter 3
A Fuzzy-Based Discounts Recommender System for Public Tax Payment

Jaime Meza, Luis Terán, and Martha Tomalá

3.1 Introduction

Taxes are the most common source of government incomes, which has influence in achieving socio-economic, political, and macroeconomic objectives of countries and other types of territorial divisions. Taxes are a legal instrument for increasing resources into the government to enhance its economic development. In the work of [25], the author highlights that tax payments are a major source of income of governments and are considered a fiscal instrument for regulating and resolving economic and social policies. At the same time, taxes are considered a mechanism for enhancing economic growth. In the case of this work, the tax payment historical behavior in Ecuador is analyzed. In Ecuador the tax payments are fundamental to financing the government national. Unfortunately, Ecuadorians' tax payment culture can be consider as low with high levels of evasion in different sectors of the economy [3, 31]. However, in spite of historical citizens' behaviors, nowadays; cultural conditions seems to be changing. According to statistics reports of the Internal Rents Service (from Spanish, Servicio de Rentas Internas[1] (SRI)), in the

[1] SRI: http://www.sri.gob.ec/web/guest/estadisticas-generales-de-recaudacion.

J. Meza (✉)
Universidad Técnica de Manabí (UTM), Portoviejo, Ecuador

L. Terán
University of Fribourg, Fribourg, Switzerland

Universidad de las Fuerzas Armadas ESPE, Sangolquí, Ecuador
e-mail: luis.teran@unifr.ch

M. Tomalá
Municipality of District of Quito (MDMQ), Quito, Ecuador

A. Meier et al. (eds.), *Applying Fuzzy Logic for the Digital Economy and Society*, Fuzzy Management Methods, https://doi.org/10.1007/978-3-030-03368-2_3

last years (2005–2016), the incomes ratio in Ecuador has been increased by 29%. This growth rate is due to public policies in taxes matter in Ecuador. Several of these policies are related to applying information and communications technologies (ICT) tools within taxes processes.

According to the European Commission[19], *eGovernment* is defined as a way to provide a wide variety of benefits including efficiency and savings for governments and businesses, transparency, and improvement of citizens' participation in political issues. One of the main challenges for the development of *eGovernment* is related to the implementation of ICT. However, *eGovernment* is much more than just ICT, it also involves rethinking organizations and processes and changing the behavior of public administrators so that public services can be delivered more efficiently to citizens.

eCommerce emerged due to the influence of the ICT revolution. Together with *eCommerce*, RSs were introduced as a mechanism to increase business incomes [35]. In the same way, several researchers have pointed out the benefits of applying the best practices of RSs to *eGovernment* solutions [26, 27, 38]. In spite of the efforts presented, the use of RSs as way to increase citizens' awareness towards improving tax payments is an application that needs more attention.

The Municipality of Quito (from Spanish, Distrito Metrolopitano de Quio (DMQ)) faces problems with a past due portfolio that includes a big group of citizens. According to report DAI-AI-0026-2017 presented by the Comptroller General of Ecuador, the past due portfolio until December 2015 was 31,284.41092 USD [10]. This trend continues generating delays in the execution of public services. On the other hand, the Municipality of Quito does not have processes and systems to setup discounts focused on citizens. Discount methods are mainly designed regarding payment deadlines. This has become in a big problem for the city. Moreover, every 2 years the most important taxes (urban and rural) are constantly changing. This process is developed and focuses on construction but it is not applied to citizens. This model request several legal reforms to change the discount formula. The constant modifications in the laws have an negative impact in the trust of citizens and it reduces their engagement level with the municipality [13, 14, 17].

These issues pushed tax authorities inside the Municipality of Quito to apply alternative and creatives approaches to reduce and mitigate the effect of a past due portfolio and motivate citizens towards improving their payment behavior [15, 16, 33]. Applying personalized discounts centered on citizen payment behavior with the use of RSs within the Municipality of Quito, could help to improve the issues mentioned above. The proposed methods in this chapter could present the following benefits:

- Improving income taxes.
- Reducing the due portfolio.
- Supporting decision-making processes to create/update taxes.
- Increasing citizens' engagement.
- Fostering the usefulness of RSs as a way to develop eGovernment.

Next sections are structured as follows. First, Sect. 3.2 gives a brief introduction and description of tax payments, RSs for *eCommerce* and *eGovernment*, and fuzzy logic, which are the basis of the model proposed. Then, Sect. 3.3 presents the methods used by the recommendation system model proposed. Additionally, in Sect. 3.4 a simulation model is presented. Finally, the authors give their concluding remarks and the future outlook in Sect. 3.5.

3.2 Background

3.2.1 *Tax Payments in Ecuador*

According to the Organization for Economic Cooperation and Development (OECD), the term *taxes* should be understood as compulsory, unrequited payments to general government, which means that the benefits provided by governments to taxpayers are not normally in proportion to their payments [30].

The current tax payment system in Ecuador is based on the following principles: legality, generality, equality, proportionality, and non-retroactivity [36]. The Ecuadorian tax system is based on indirect and direct taxes. In 2015, the fiscal pressure in Ecuador was around 21.7% [9]. On average, taxes represented 76% of the Ecuadorian GDP in 2016 [7] and are administrated by the SRI. On the other hand, external taxes are administrated by the National Customs Service of Ecuador (from Spanish, Servicio Nacional de Aduana del Ecuador[2] (SENAE)), which also has nationwide jurisdiction. Local and provincial taxes are created and administered by each local government as municipalities and provincial councils, respectively.

Municipalities in Ecuador use different discount methods as way to persuading citizens to pay their taxes. The most common method is to offer discounts by early payment [5].

3.2.2 *RSs for eCommerce and eGovernment*

RS techniques are applied in several domains. In the same way, Schafer et al. [35] show evidence that RSs can increase businesses' incomes. Additionally, the survey developed by Lu et al. [27] presents the state of the art in recommender system applications. A list of domains were evidenced with positive outcomes, which includes fields such as: *eGovernment*, *eCommerce*, *eLibrary*, *eLearning*, *eBusiness*, *eTourism*, *eResources*, and *eGroups*, among others.

In the academic literature, the most used mechanisms to develop RS applications are: collaborative filtering (CF), content-based (CB), and hybrid [8]. The main

[2]SENAE: https://www.aduana.gob.ec/.

problem in the CF approach is related to the so-called sparsity problem, when the number of items rated is small compared to the total number of items [38]. On the other hand, CB faces the problem of overspecialized recommendations [27]. To solve these problems, many advanced recommendation approaches have been proposed including social network-based RSs, fuzzy-based RSs, context awareness RSs, and group RSs [27].

In the work of Yager [41], the author points out that an important component of CF that is the calculation of similarity of interest based on correlations between individuals for predictions and recommendations. Several issues have been identified when using CF approaches, based on this issue. For that reason, a new approach so-called reclusive approach, was proposed. It is based upon finding a similarity between objects while the CF approach is based on similarity between people. This approach considers an RS as a collection of objects as, $O = \{o_1, \ldots, o_n\}$ and it recommends to a user objects of O that could interest him. In [41], the proposed method considers five issues: RS as object collections, object representation, user preferences modeling, user profiles, and environment.

The basis of this approach is an object and its membership degree. According to Zadeh [43], an object do not have a defined membership criteria. For instance, in the case of the class *animal*, it has a clear definition (e.g., dog, cat, horse, etc.). However, the class *people's ages* (e.g., baby, boy, young, and old), an imprecise decision emerges. In this arena, where the decision-making process is not obvious, fuzzy sets and fuzzy logic appears as a solution.

3.2.3 Fuzzy Logic Overview

In the work of Zadeh [43], the author pointed out that fuzzy sets provide a natural way of dealing with problems in which the source is imprecision. On the other hand, in sharp logic, the values obtained after executing a sentence are binary (true, false). These results are represented by *crisp sets*; however, several solutions in real life need values in the interval [0, 1]. In this case, fuzzy logic is similar to human decision-making with its ability to work with approximated data to find precise solutions [1].

Fuzzy set theory was introduced by Zadeh in 1965 [43]. It described the concept of membership degree. *Crisp sets* behaviors are denoted by: If x belongs to a set A then $x \in A$, otherwise $x \notin A$. Therefore, for each x of the set A there are only two responses: either x is an element of A or it is not. Using the concept of membership degree, each element can be represented by a function that defines the values allowed. The values allowed are drawn as linguistic values. A linguistic value refers to a label representing knowledge that has meaning determined by its degree of membership function. For instance, $x_1 = old$ with the degree $\mu = 0.8$ means that the variable x_1 has a linguistic value represented by the label *old*, whose meaning is determined by the degree 0.8.

To illustrate this concept in the case presented in this chapter, a *crisp set* for the linguistic variable $discount$ as a function of the payment date of a citizen is used. It tries to resolve the following problem: A local government institution wants to create a discount over one of its taxes. The discount strategy expects that citizens pay as soon as possible to maximize their annual budget. In represents that if one citizen pays between the discount period he gets a discount, but, if he pays in the first days he gets a gradual discount.

The class *discount* storage the discount periods in days. It is defined as follows: $discount = \{1, .., 180\}$. The characteristic function presented in Eq. (3.1) has only two values: true (ten) or false (zero).

$$X_{discount}(x) = \begin{cases} \text{if } 1 \leq x \leq 180 & 1 \\ \text{if } 180 < x < 366 & 0 \end{cases} \tag{3.1}$$

The function presented in Eq. (3.1) is not useful for the calculation required by the local government institution, since it does not allow setting a real discount value according to payment date.

This issue can be resolved using fuzzy sets with membership functions. Zadeh gives a number for each value inside the universe set; this number is the degree in which the element is inside the set. For instance, in the case of the set *discount*, there are some values that represent the citizen payment date; therefore, different degrees of *discount* could be represented.

Zadeh defines fuzzy sets as A in X is characterized by a membership function $f_A(x)$ that associates each element in X with a real number in the interval $[0, 1]$, with the value of $f_A(x)$ at x representing the grade of membership of x in A. Equation (3.2) shows the membership function for the fuzzy class *discount*. The equation presented in Eq. (3.2) responds to a gamma inverse fuzzy function definition.

$$X_{discount}(x) = \begin{cases} \text{if } 180 < x < 366 & 0 \\ \text{if } x \in (1, 180) & 1 - ((x-1)/(180-1)) \\ \text{if } x = 0 & 10 \end{cases} \tag{3.2}$$

The level of granularity used in Eq. (3.2) allows the local government institution to apply a real discount related on payment date. The real values after and before could be represented in linguistic variables by a collection of quantifiers such as: near, close, approximately, etc. [44]. Finally, fuzzy set theory uses a large volume of operations and properties that can be considered as tools to apply in different scenarios.

3.2.4 Fuzzy Logic Applied to Marketing

In the work of Donzé and Meier [18], the authors define marketing as way to identify and pick up the customers' needs. They also point out the use of customer relationship management (CRM) as a strategy for building customer equity and improving financial revenue. In this sense, fuzzy logic can be applied to marketing in a wide set of tasks.

eGovernment is defined as a subset of the exchange relationships of *eCommerce*. Therefore, *eMarketing* is a way to improve the relations between citizens and government institutions through ICT tools. In the work of Meier and Stormer [28], authors propose to add *eMarketing* as an element into the value chain of *eBusiness*.

Several researchers highlighted the importance of fuzzy logic in both *eCommerce* and *eGovernment* as an effective means of decision support, not only from a business perspective but also from the clients or citizens perspective. Nowadays, fuzzy classification models are used in the implementation of RSs. In the work of Lu [27], the author highlighted a number of RS approaches and the techniques implemented. A summary is presented in Table 3.1.

Additionally, other authors have presented positive outcomes on the use of RSs in *eCommerce* and *eGovernment* [11, 12, 22, 23, 26, 27, 39, 42]; however, the application of RSs as a way to increase the citizens' awareness towards tax payments is poorly explored.

Table 3.1 RSs techniques applied, adapted from [27]

	Feature		
Name	*Domain*	*Technique*	*Period*
Smart participation [37]	*eGovernment*	Fuzzy clustering	2014
TPLUFIB-WEB	*eGovernment*	Fuzzy linguistic modeling, hybrid, CB, CF	2014
Methods for therapy [21]	*eHealth*	CF, Demographic-based RS	2016
Smart BizSeeker [27]	*eGovernment*	CF, hybrid, fuzzy sets	2013
Procurement [45]	*eGovernment*	Fuzzy logic, item-based CF, Bayesian approach	2015
Dissemination of information in university digital libraries[32]	*eLibrary*	Hybrid, 2-tuple fuzzy linguistic approach	2017
Proactive and reactive e-government services[6]	*eGovernment*	Hybrid with ontology-based recommendation model	2015
RS for elderly people [34]	*eHealth*	Hybrid	2015
Personalized E-learning [40]	*eLearning*	Hybrid, fuzzy tree and learner profile	2015

3.3 Fuzzy-Based RS Model

The model presented in this chapter is centered on citizens and their payments behaviour. Therefore, the reclusive approach for recommendation presented by Yager [41] is applied.

Based on the targeted marketing definition as results of RSs in *eCommerce* field, the approach proposed uses the definition of *discount* as a way to persuading citizen to paying taxes. In this way, it creates personalized discounts to each citizen. The model is set by two types of discounts centred on citizens, general and specific, defined as follows.

- General discount—It is applied in relation to the historical payment behaviour.
- Specific discount—It is related to specific issues that e-government institutions could be interested (e.g., dead line, risk relevance, etc.)

Both, general and specific discounts have been implemented using fuzzy sets definitions. Such definitions were mixed in a Euclidean space to set specific discount for specific citizens, where each discount percentage is unlikely. In the next section, the fuzzy sets used and its details are shown.

3.3.1 Fuzzy Sets

This section aims is to shows every fuzzy definition used to implement the discount methods but also the recommendation notification policies.

3.3.1.1 General Discount

To implement the general discount, the proposed model uses the historical citizens behaviors considering the number of times that the citizen was included in some payment condition as it is show in Table 3.2.

Payment conditions are applied regarding a group of taxes in the past. The values are computed using the fuzzy classification sets presented in Fig. 3.1. The

Table 3.2 Condition of payment by taxes

	Feature		
Condition	Rating (R)	Lexical meaning	Target taxes
Discount payment	+++++	Excellent	All
Non-discounted payment	++++	Good	All
Penalty payment	+++	Average	All
Payment after notification	++	Regular	All
Payment after judicial notice	+	Bad	All

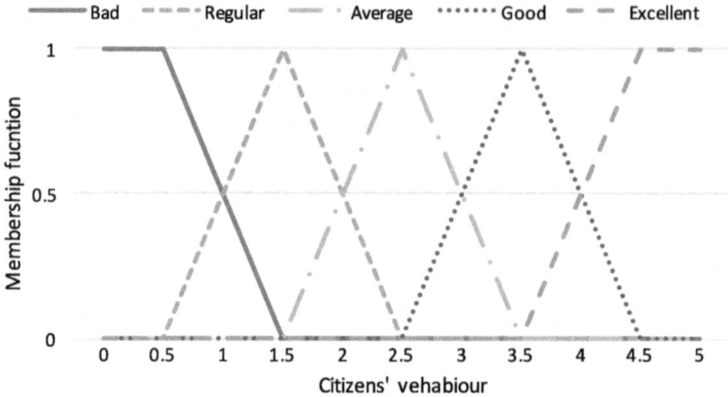

Fig. 3.1 Fuzzy sets for citizens behavior

Table 3.3 Discount by rating

Linguistic value	Discount %	Rating
Excellent	81–100	3.5–5.0
Good	61–80	2.5–4.5
Average	41–60	1.5–3.5
Regular	21–40	0.5–2.5
Bad	0–20	0–1.5

combination presented in Fig. 3.1 allows us to set a citizens classification, as well as a membership degree, defined as citizen behavior ranking (CBR).

Five fuzzy sets were drawn (Fig. 3.1) over the linguistic variable *citizens behavior*. The linguistic values are related to the citizens payment behavior (i.e., bad, regular, average, good, and excellent). Each linguistic values have a corresponding discount by crisp group as shown in Table 3.3. The ranking values shown in Table 3.3 have a corresponding fuzzy set (refer to Fig. 3.1); however, each ranking value is overlapped by other value. Therefore, the discount percentage does not belong to an element in one category only. In this scenario, the final discount will be done by membership degree on each group of fuzzy sets and the range of discounts set as shown in Table 3.3. This result is named citizen behavior discount percentage ($CBDP$) as shown in Eq. (3.3).

$$CBDP = \frac{\sum_{i=1}^{n} \{|(RF - (RI - 1))| \times \mu_i\}}{n} \tag{3.3}$$

where RF is the final value on the column discount in % as shown in Table 3.3. RI is the initial value on the column discount in % in Table 3.3. μ_i is the membership degree of the fuzzy set i. Finally, n numbers of fuzzy sets overlapped.

$CBDP$ computes the discount percentage by every group of taxes to compute the global percentage discount by citizen ($GCBDP$). Each group of taxes is assigned a weight (W) as shown in Eq. (3.4).

$$GCBDP = \sum_{I=1}^{N} W * CBDP \tag{3.4}$$

where, N is the total number of group of taxes by citizen. W is the ratio between the value to paying of tax group and the total debt for the citizen. As an example, citizen A has to pay three taxes: urban, patent, and operation. His total debt is 100 and the values to pay by each group of taxes are 80, 10, and 10. Therefore, the value of W assigned to each group of taxes are: $urban = 0.8$; $patent = 0.1$, and $operation = 0.1$. Finally, $CBDP$ is the individual discount percentage by group of taxes.

3.3.1.2 Specific Discount

The general discount presented in the previous section is the key element in the citizens payment behaviour. It could be mixed with specific discounts related with other factors (e.g., dead line, risk relevance, etc.). Specific discounts percentages (SDP) are computed as ratio between a point inside of Euclidean space and the maximum distance value (MDV) that can take negatives or positives values of the scale. It is shown in Eq. (3.5).

$$SDP = \frac{\sqrt{\sum_{i=1}^{n}(x_i - y_i)^2)}}{MDV} \tag{3.5}$$

Both general and specific discounts are added for getting the final discount (FD) by group of taxes. It is shown in Eq. (3.6).

$$FD = \frac{(CBDP + SDP)}{2} \tag{3.6}$$

According to previous definitions, the model proposes two specific discounts. The first one is related to deadline payment, and the other with the risk relevance. Both discounts are combined with CBR and plotted on an Euclidean space as shown in Fig. 3.2. To compute the discounts according to Fig. 3.2, consider the X axis in region dead line represents time, and risk relevance is the ratio between citizens debt and the citizens with the most debt inside each group of taxes; moreover, points labeled as $X1 - Y1$ (as shown in Fig. 3.2) are the initial points to compute the Euclidean distance.

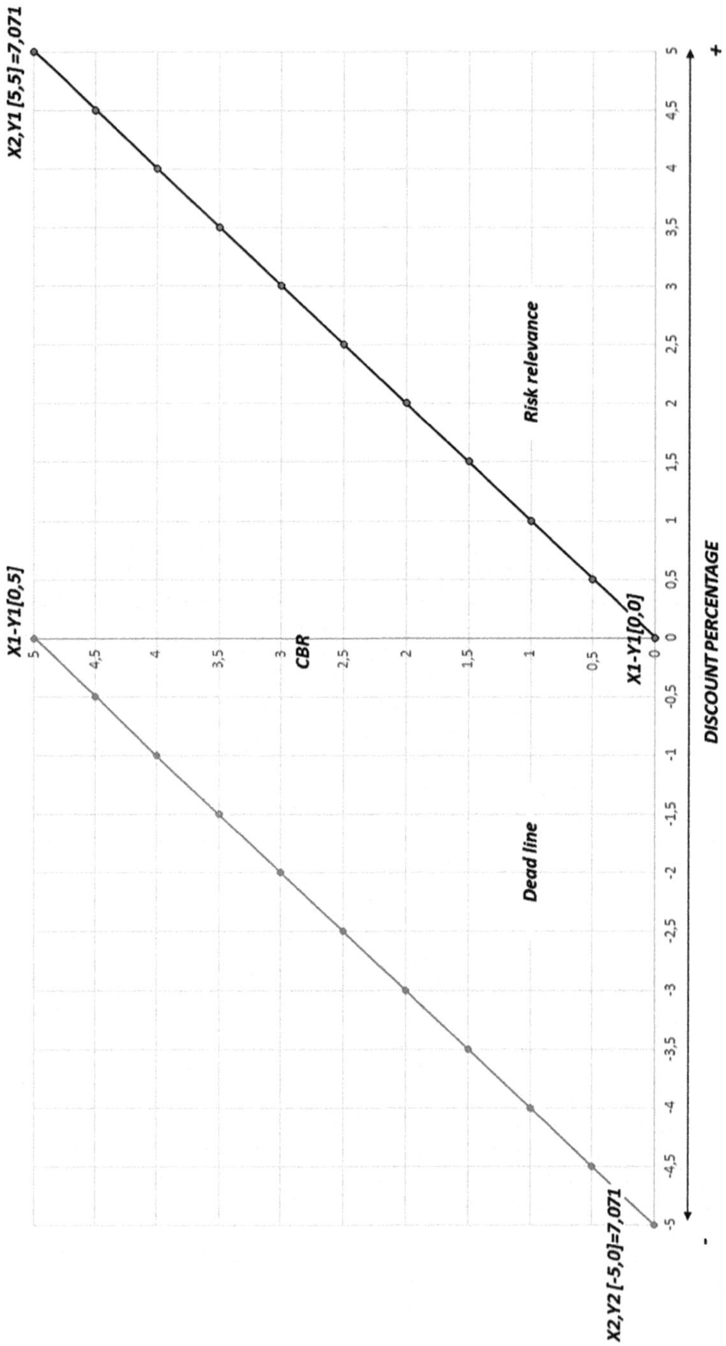

Fig. 3.2 Discounts combined with CBR

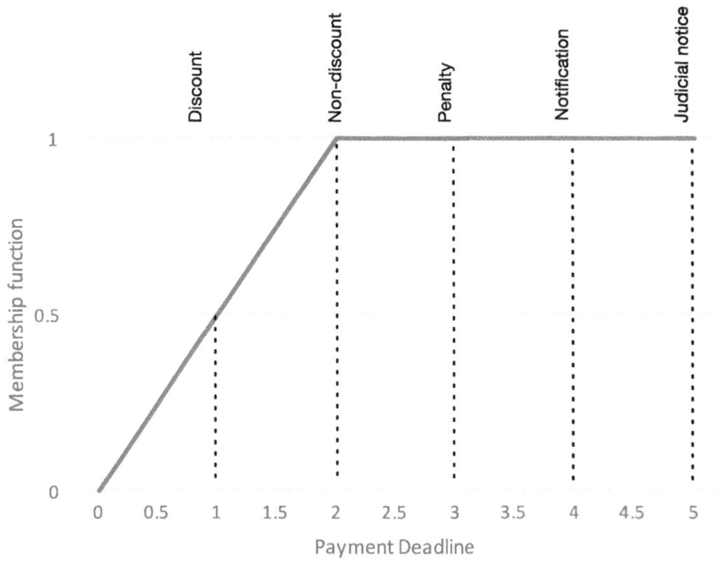

Fig. 3.3 Fuzzy sets by payment deadline

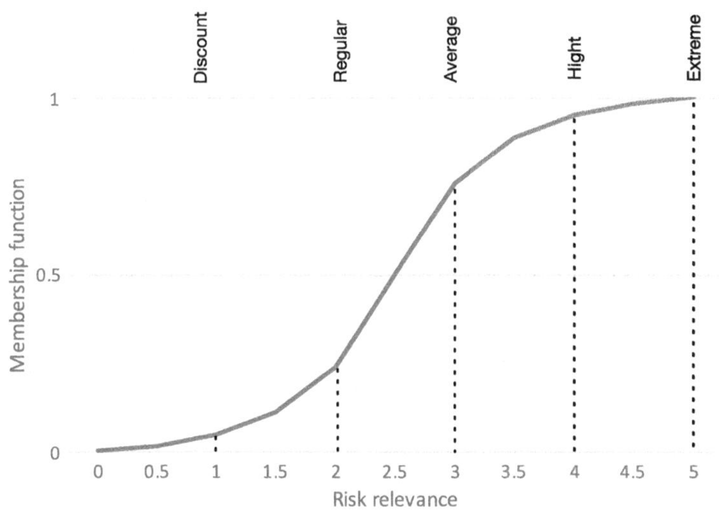

Fig. 3.4 Fuzzy sets by risk relevance

3.3.1.3 Recommendation Notification Policies

The fuzzy sets shown in Figs. 3.3 and 3.4 allow to setup the membership degree to define different recommendation policies, for example, if the membership function is 1 in both scenarios the recommendation should be sent to citizen every standard time.

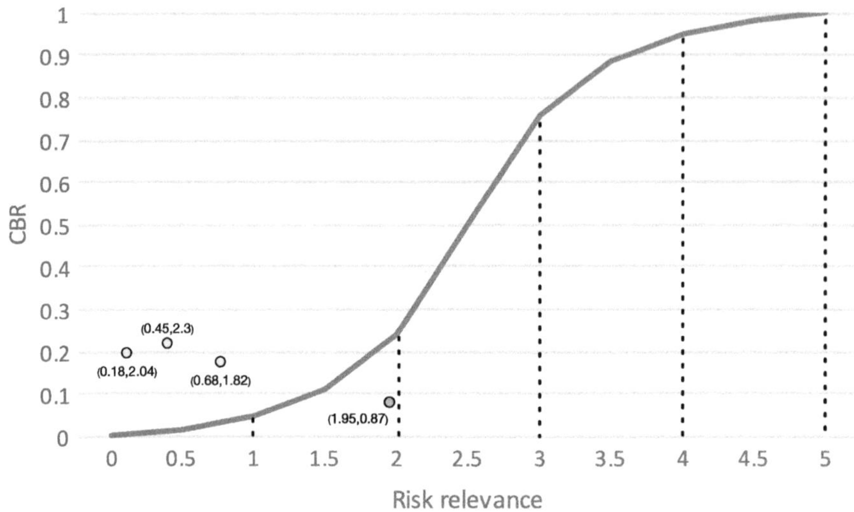

Fig. 3.5 Fuzzy profile (FP)

Citizens behavior fuzzy sets allow establishing recommendations centered on citizen. Therefore, the proposed model combines every fuzzy set (payment deadline and risk relevance) in an Euclidean space to compute the centroid defined as fuzzy profile (FP). Figure 3.5 presents the fuzzy sets of the citizens behavior and risk relevance combined in an Euclidean space, it is computed using Eq. (3.7). The FP is computed by setting up the initial centers as $(0, 0)$. The X axis represents the relevance (or deadline) and the Y axis is the point generated by CBR as shown in Fig. 3.5.

$$FP = \sqrt{\sum_{i=1}^{n}(x_i - y_i)^2} \tag{3.7}$$

The FP presented in Fig. 3.5 generates a recommendation point with value 0.87. The risk relevance is calculated considering the value five for the citizen with the highest tax debt; therefore, the relevance for a specific citizen is the ratio between his or her debt and the citizen with the highest debt.

This information changes in *real time for each citizens payment.* On the other hand, the deadline for payment relevance is calculated considering the value two (non-discounted payment) as the user is moving towards the risk zone. Therefore, in this case notifications will be send. The notifications regarding citizens payments will depend on their behaviors. Both risk relevance and deadline payment generate a *recommendations matrix* to save recommendations messages.

3.3.2 System Architecture

The system architecture is show in Fig. 3.6. It considers three components: message handler (MH), request validation (RV), and fuzzy recommender system (FRS).

3.3.2.1 Message Handler

The message handler (MH) implements several functions to send recommendations available to either citizens or government agencies in different ways, for example, e-mail, SMS, voice mail, etc. The recommendation message is splitting according to the fuzzy sets by payment deadline, risk relevance, and citizen behavior. For each citizen C_i, a set of recommendation R_i will be send. Each R_i could be responded by the citizen; therefore, the ranking of accuracy of recommendations is computed by the ratio of response sent for each recommendation. Responses are received by a request validation (RV) component.

3.3.2.2 Request Validation

The RV implements a reactive autonomous agent as a real time supervisor a shown in Fig. 3.7. Its main function is to listen the environment, activate the process corresponding and increase the knowledge base. Figure 3.7 presents the inputs and output from this component.

The RV component is spitted by six inputs, three outputs, and six processes to control the states and one data base (refer to Table 3.4). Each input is considered as a sensor or listener, its waiting triggers or shift in the environment are designed to fire an action. The outputs are responsible to connect the message or order from the RV towards its target. The processes have to comply every rule or restriction until they get a successfully output requested. Finally, the database records each interaction and its results.

3.3.2.3 Fuzzy Recommender System

The fuzzy recommender system is divided by three sub components, the central data repository (CDR), the constraint-based engine (CBE), and the fuzzy recommendation engine (FRE).

Central Data Repository

Institutional data warehouses are widely applied in e-government solutions [2, 4, 29]. Therefore, Central Data Repository CDR implements an institutional data warehouse (DWH). Its main function is to integrate the information from large number of homogeneous and/or heterogeneous sources into an unique data storage.

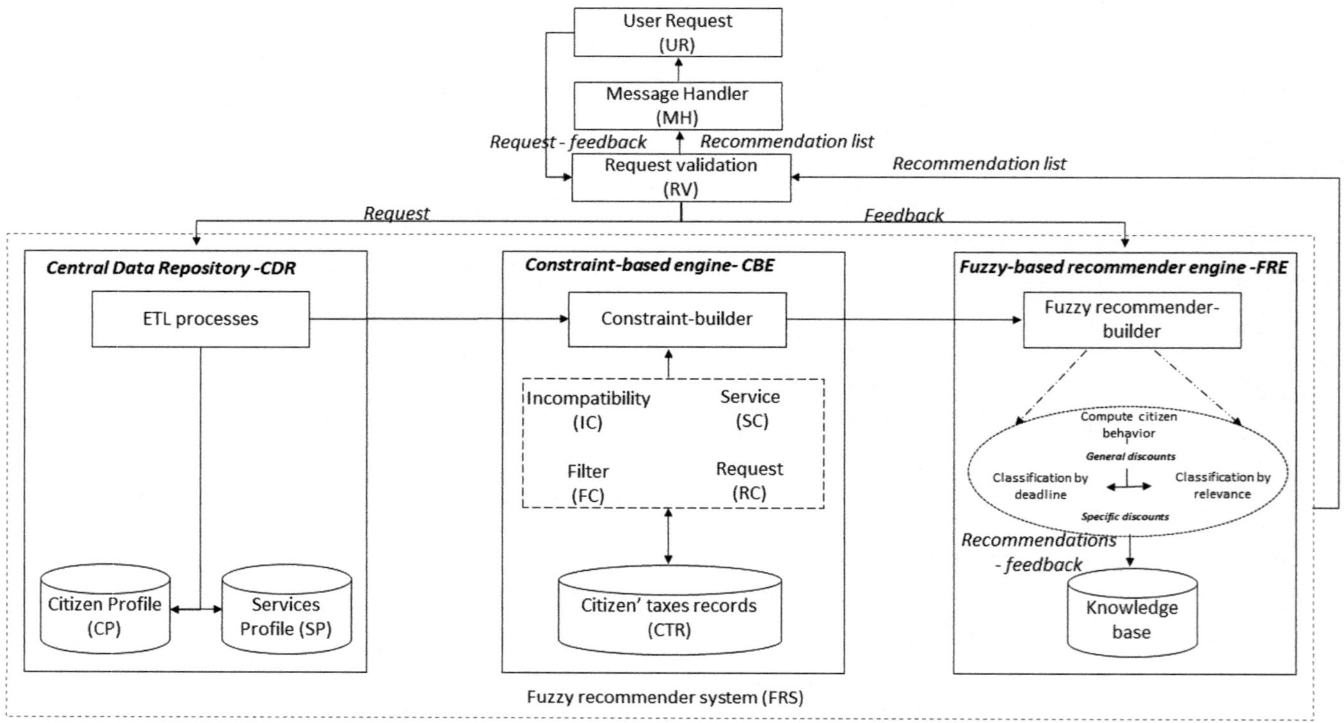

Fig. 3.6 Fuzzy-based recommender system architecture

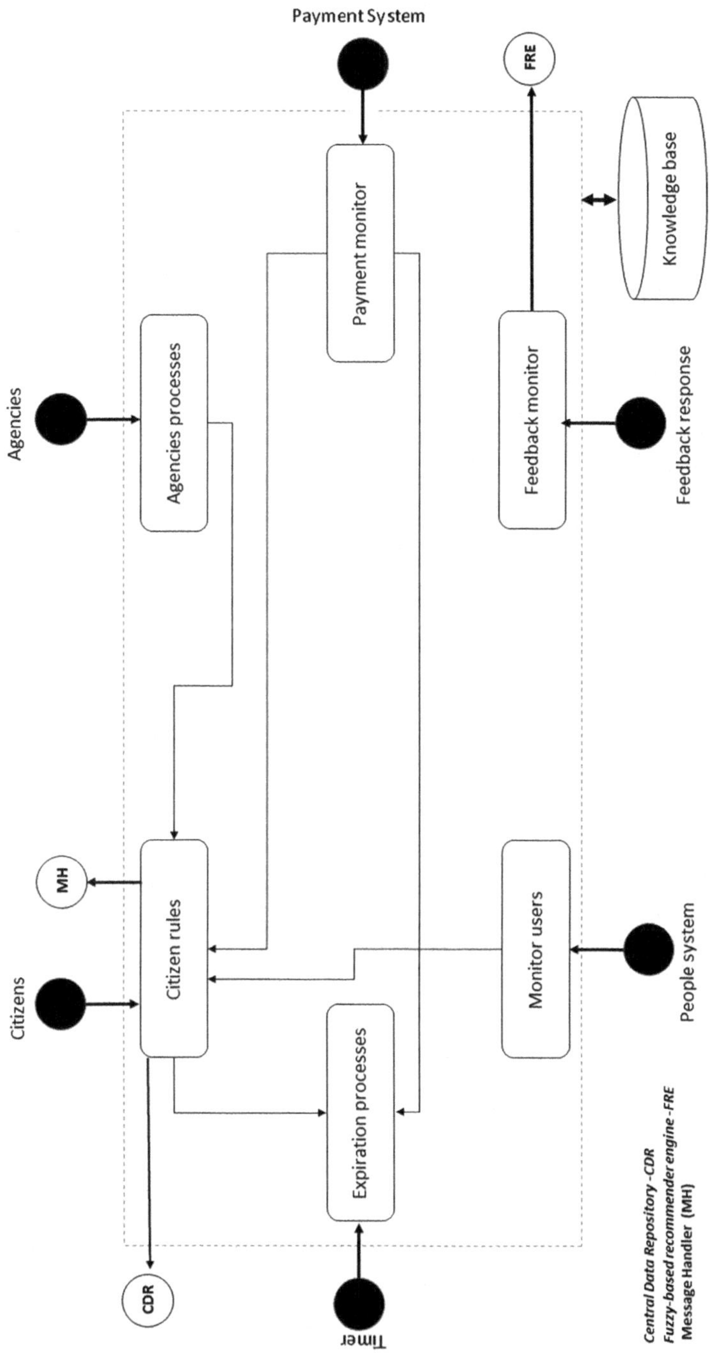

Fig. 3.7 Request validation design

Table 3.4 Elements of the request validation component

Kind	Name	Roles
Input	Citizen	It catches the actions from users registered in the website called PAM http://pam.quito.gob.ec.
Input	Agencies	It is waiting for an order from the agency operator, order such as taxes reports, recalculation of taxes, and resend recommendations
Input	Payment system	It is triggered every time that a user execute a payment by any way (e.g., credit card, banks or local cash desk)
Input	Timer	It monitors continuously the state of recommendations expiration date and throws the instructions an order to regenerate the recommendation list
Input	People system	Every time that a citizen is enrolled inside of the system it executes the recommendations related to the new citizen or data updated
Input	Feedback response	It caches every feedback from citizens related to recommendations
Output	FRE	Allows updating recommendations feedback from the citizen
Output	CDR	It starts the processes to compute the discount and recommendations
Output	MH	Entrance to recommendation lists, if needed to be sent to citizens
Processes	Citizen rules	It executes every constraints related to citizens such as notifications, change of relevance, and payment condition improved
Processes	Agencies processes	Turns on the process by specific demand
Processes	Expiration processes	It executes triggers to check the state of recommendations expiration state
Processes	Payment monitor	It connects payment services with the need to send recommendations
Processes	Monitor users	It connects payment services with need of send recommendations
Processes	Feedback monitor	It updates the recommendation with feedback into the knowledge base
Data base	Knowledge base	It records every action over recommendation and the discount generated

Moreover, the DWH has different levels of details, one of this is defined as *data mart* [24]. Data flows into the DWH from the environment which needs processes to load and transform the data, these processes are named *extraction transformation and load* (ETL). As shown in Fig. 3.7, CDR contains ETL' processes to process data from citizens and services. It's more important features are summarized in Table 3.5.

The CPD stores the profiles of involved citizens. A citizen profile (CP_i) is associated with a citizen C_i. It consists of a tuple ($CPDF_i$, $CPDP_i$, $CPDT_i$), where:

Table 3.5 Central data repository features

Name	Description	Data stored	Update mode—frequency
Citizens (CPD)	Citizen profiles and preferences	Features ($CPDF$), preferences ($CPDP$), and taxes ($CPDT$)	ETL—daily/by demand
Services (SPD)	Municipal services features	Features (SPDF), budget (SPDB), and constraints (SPDC)	ETL—by week/by demand

1. $CPDF_i$ stores the identification, social, and demographics information such as, name, date of born, gender, province, zone and so on;
2. $CPDP_i$ infers the rules from the number of times that the services (taxes) were used in the past;
3. $CPDT_i$ stores information of citizens account about pays and debts (statement of account). It includes information such as, payment date, value of pay, agency, discount.

The SPD stores the profiles of involved *eServices*. A services profile SP_i, associated with an service S_i consists of a tuple $(SPDF_i, SPDB_i, SPDC_i)$, where:

1. $SPDF_i$ stores the identification, descriptions, and demographics information such as, name, date of deliver, target, zone and so on;
2. $SPDB_i$ stores the information of service budget including the cost of services without an electronic service, the cost using electronic service, and quantity citizens save using electronic services, among others;
3. $SPDC_i$ stores the information of constraints in the service such as, age, yearly income, nationality, permanent residence, physical disability, mental disability, job injury, qualification, academic degree, marital status, number of children, and relevance, among others.

The column "update mode frequency" presents the time executed by the ETL processes. Every data repository (CPD) or (SPD) are executed one time per day or according to requests made from external agents.

Constraint-Based Engine

Constraint-based Engine CBE was adapted from the constraint-based recommender systems approach in [20], and the framework proposed in [6]. Constraint-based recommender systems approach recommends services according to citizens needs [20]. In order to adapt the approach previously mentioned in the case of taxation on eGovernment services, a set of variables related to citizens profiles, services profiles, and a set of constraints are used. They define the relation between them, this set of constraints are named "Constrains-builder". Its content is presented as follows:

1. Incompatibility (IC) restricts the services according to citizens requests (e.g., age, incomes, etc.)

2. Service (SC) owns restrictions of the service (e.g., availability, date launched, budget, etc.))
3. Filter (FC) relates between citizens request and services (e.g., business taxes, urban taxes, etc.)
4. Request (RC) constraints related citizens requirements and services assigned (e.g., citizen address).

Information of relation between Citizen Profile and Service Profile in the matter of taxation is saved in Citizen Taxes Records data source (CTR). Data available on (CTR) contain information such as: Tax , citizen , payment date, value to pay, value paid.

Fuzzy-Based Recommender Engine

Fuzzy-based Recommender Engine (FRE) processes the data available on the citizen tax records data source (CTR) and prepares the recommendation list. The recommendation engine components presented in Fig. 3.6 implements the rules mentioned in Sect. 3.3.1. Recommendation lists are prepared using a set of steps defined in the fuzzy recommender builder presented as follows.

1. Citizen behaviour. It computes the citizens behavior by rating them from historical payment data. Every payment is assigned with a rating according to Table 3.2. The final rating is the average of the data set processed.
2. General Discount. With the citizens behaviour rating, the general discount percentage is computed as a membership degree by each group of fuzzy sets multiply discount range as presented in Table 3.3 and Eq. (3.3). As an example, a user rated with 3.2, his discount will be computed as membership degree of fuzzy sets *good* ($\mu_G(3.2)$) and average ($\mu_A(3.2)$) with the range of percentage settings in Table 3.3.
3. Computed classification by deadline. It computes the membership degree in the fuzzy set as shown in Fig. 3.3. Citizens with membership degree greater than 0 should be sent a paying recommendation.
4. Computed classification by relevance. It computes the membership degree in the fuzzy set as shown in Fig. 3.4. Citizens with membership degree greater than 0 should be sent a paying recommendation.
5. Computed specific discounts. It is used to compute specific discounts. The fuzzy recommender-builder implements the rules mentioned in Sect. 3.3.1.2.
6. Ranking. It represents a combination of sort processes. It implements three rankings, by discount values, risk relevance, and citizens behavior.
7. List of recommendations. The list of recommendations are performed after the classification calculation by relevance and deadline. Its responsibility is sending details by every recommendation towards the request validation (RV) component and updates the knowledge base with the new information generated.
8. Feedback. It updates the knowledge base with information responded by the recommendation sent.

Table 3.6 Land taxes restrictions

Type	Restriction	Evaluation	Approved
Incompatibility	Citizen age is higher than 18	Citizen age is 28 years old	Yes
Service	Law for land taxes in 2018 is approved	Law approved on December, 2017	Yes
Filter	Citizen is owner of apartment, house or land space	Citizen has a suite	Yes
Request	Citizen has pending invoices in order to service requested	Citizen does not pay deb in 2018	Yes

3.3.3 *Recommendation Computation Process*

In this section the case of the citizen of Quito city is used to follow the recommendation process. As an example, the case of the citizen Juan is used. He is 28 years old, lives in Quito in his suite, in the last 5 years he paid his municipal taxes as follows, three times with discount of 8% and two times with punish of 3%. In the present year Juan needs more information about the best way to save money. Juan makes this request on March 5, 2018 and he has a debt of USD 125.20. Juan has an account in municipal website PAM.[3] The maximum debt inside of the group of tax when Juan sent the request was USD 2125.00. The municipality has set in 2018 as max discount of 10%. From January until June, the municipality offers a discount by early payment.

According to the fuzzy-based recommender system architecture presented in Fig. 3.6, the steps to follow are:

1. Request is processed by (RV) as "Citizen" and executes "Citizen rule" process (Fig. 3.7), then a request is accepted and sent to the FRS given that Juan has an account in the municipal website.
2. Juan has data from previous year; therefore, the ETL process is skipped.
3. A set of restrictions are applied by the "Constraint-builder". Those restrictions are evaluated and the request approved as shown in Table 3.6. The data is not update in the "Citizen' taxes records", since their were not changed.
4. Juan historical payment behaviour is computed as rating of 4.2; therefore, Juan is considered as citizen between *Good and Excellent* (refer to Fig. 3.1). The rating assigned to Juan allows him to get a general discount of 37.5% (refer to Eq. (3.3)) over the max allowed by the municipality (10%). That's means a real general discount of 3.75%.

[3]PAM: http://pam.quito.gob.ec.

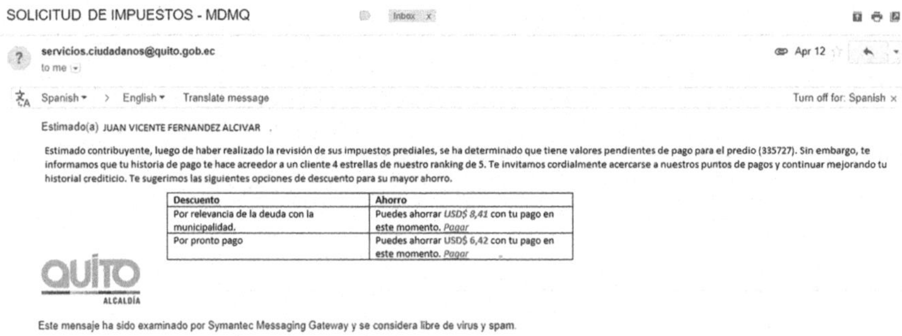

SOLICITUD DE IMPUESTOS - MDMQ Inbox x

servicios.ciudadanos@quito.gob.ec
to me
Apr 12

Spanish > English Translate message Turn off for: Spanish ×

Estimado(a) JUAN VICENTE FERNANDEZ ALCIVAR .

Estimado contribuyente, luego de haber realizado la revisión de sus impuestos prediales, se ha determinado que tiene valores pendientes de pago para el predio (335727). Sin embargo, te informamos que tu historia de pago te hace acreedor a un cliente 4 estrellas de nuestro ranking de 5. Te invitamos cordialmente acercarse a nuestros puntos de pagos y continuar mejorando tu historial crediticio. Te sugerimos las siguientes opciones de descuento para su mayor ahorro.

Descuento	Ahorro
Por relevancia de la deuda con la municipalidad.	Puedes ahorrar *USD$ 8,41* con tu pago en este momento. *Pagar*
Por pronto pago	Puedes ahorrar USD$ 6,42 con tu pago en este momento. *Pagar*

QUITO
ALCALDÍA

Este mensaje ha sido examinado por Symantec Messaging Gateway y se considera libre de virus y spam.

Fig. 3.8 Screenshot citizen recommendation message

5. Juan does not pay yet; therefore, he will be notified by deadline. On the other hand, he will not be notified by risk relevance, since his rating of behavior is 4.2 and his ratio of debt is 5.89%. They are not considered as relevant for early payment.
6. Juan percentage specific discounts are: deadline = 13.79%, risk relevance = 29.70%.
7. Juan will be informed with the suggested list to paying as shown in Fig. 3.8.

3.4 Model Simulation

This section evaluates the outcomes of the proposed model using a sample from the Municipality of Quito data set. The model simulation aims are:

1. To evaluate citizens historical payment behavior
2. To verifying the data set behavior with the discount model proposed
3. To summarize the outcomes over the data simulated

This section is organized as follow: in the first part the data set structure is analyzed, after that the simulation design is presented, and finally, some outcomes are presented.

3.4.1 Dataset Acquisition

In order to prepare the simulation design, the data sets from municipality taxation systems were collected, prepared, and analyzed. The dataset used has the structure presented in the Table 3.7.

Table 3.7 Data set structure for taxes payment

Name	Type	Description
CitizenID	Numeric	Unique identifier for citizen
TaxID	Numeric	Unique identifier for services or tax
TaxGroupID	String	Unique identifier for tax group, for instance, urban or business taxes
PayCompulsoryDate	Date	Deadline by type of tax
RealPayDate	Date	Real date of pay
ValueRequested	Decimal	Value requested to pay
ValuePaid	Decimal	Real income by pay
PayCondition	Char	It stores different states for citizens according to tax as follows: 1 discount, 2 no discount, 3 penalty, 4 penalty notification, 5 penalty after judicial notice
PeriodValue	Numeric	Period of tax by year or semester

Table 3.8 Municipal taxes data set

Feature	Number	Description
Total records	10,286.201	Total number of register available for processing
Citizens processed	1135.733	Citizens that have urban or business taxes
Type of taxes	20	Type of taxes available were classified by two groups: urban and business taxes
Tax groups	2	Deadline by type of tax
Period processed	34	It contains data from period 1985–2018

Table 3.9 Kind of municipal urban taxes processed

TaxId	Name
1	Urban predial
2	Urban predial parish
3	Rural predial
4	Pavements
5	Building site
34	Urban determinations
35	Rural determinations
50	Predial recalculation year 2012–2013
51	Predial recalculation year 2014

The structure shown in Table 3.7 was filled up from citizen profiles (CP) and service profiles (SP) using an ETL processes of central data repository (CDR) (refer to Fig. 3.6). Data processed is summarized in Table 3.8.

Data sets acquisition was applied to Urban Taxes, this group considered a 194.971 citizens that had to pay urban taxes in some one of the categories presented in Table 3.9.

3.4.2 *Simulation Design*

In order to prepare scenarios of simulation models, several steps were completed, as so also tools of support were built (refer to Fig. 3.9).

3.4.3 *Outcomes*

- The average rating for citizens was 3.44/5 stars. This sample was split according to pay behaviour: discount payment 9%, non discounted payment 53%, penalty payment 35%, and payment after judicial notice 3%.
- The citizens that paid with discount their mean of discount was 5.01% and their deviation 2.91; therefore, the discounts gotten by citizens were between 2% to 8%.
- The discount approach proposed presented a mean of discount of 7.68% and it deviation 0.064, therefore, the tax discount distribution is more close between citizens.

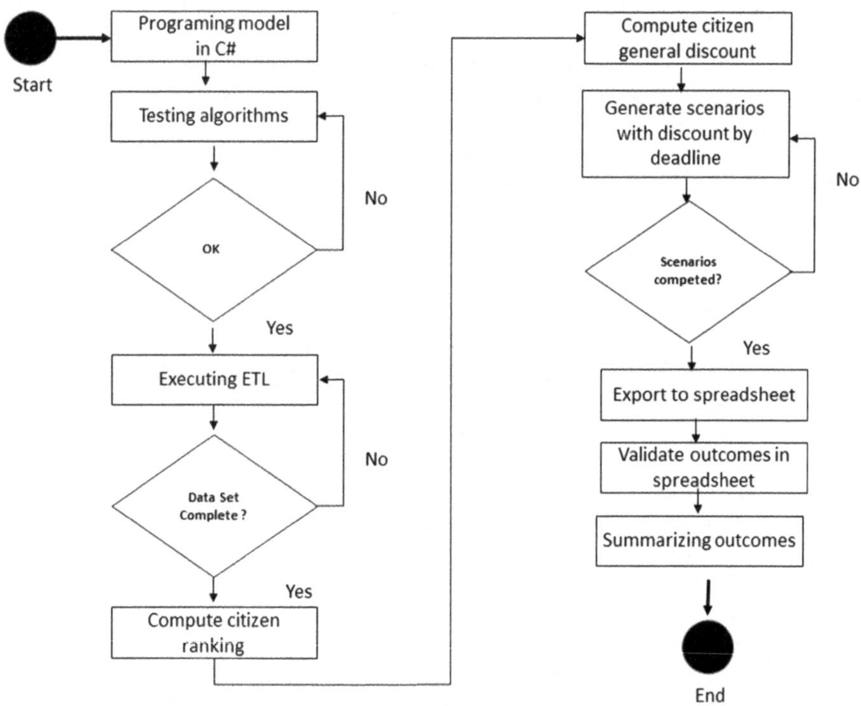

Fig. 3.9 Simulation design process

- The citizens that paid with discount in the past could get at least 4.9% of discount in future payments.
- Three scenarios were simulated to specific discount. In the worst case scenario citizens could get at least 5.98% of discount in future payments.
- The citizens that never paid with discount in the past could get at least 5.7% of discount in future payments.

Outcomes presented show evidence that the discount strategic used by the municipality was not adequate given that percentages of taxpayers that paid with discount are low. On the other hand, the discount tax model applied gives insights about its opportunities to improve municipally incomes using this strategy.

3.5 Conclusions, Limitations, and Future Work

In this work, a fuzzy-based recommender system was introduced. It was applied to notifying citizens regarding public tax payments and also discount opportunities. The proposed model has been tested using a simulation from the Municipality of Quito dataset. In this work, the model presented behaviors with a positive impact in citizens tax awareness.

This hypothesis is supported in the highlighted results of the simulations. For instance, 91% of citizens are below the level of relevance two; however, their payment behavior rating was presented inside the average. On the other hand, the rest of the citizens payment behavior ratings were consider as regular. The proposed model shows an effective way to identify and recommend this group of citizens.

The discount approach proposed in the model, show effectiveness in the way for compute taxes, therefore, that is an initial point in order to the municipality rethinking the strategies of discounts.

Another conclusion is that fuzzy logic approach applied as a classification method improves the discount balance between citizens and it gives broad options towards new discounts methods.

The recommendation approach presented in this work is centered on citizens behavior as strategy to recommender payment options to compute discounts scenarios by citizen.

Future work should focus on the design machine learning approach to improve the real time supervisor; combining both CF and reclusive approaches to propose recommendations using group preferences by finding other citizens who could have similar behaviors. The next stage for the proposed model is to test it with the citizens from the Municipality of Quito. Data gathered with citizens interactions will be used for impact evaluation and redesign of the proposed RS approach.

Acknowledgements Authors would like to thank the members of Information System Research Group (http://diuf.unifr.ch/is/) at the University of Fribourg for contributing with valuable thoughts and comments. We specially thank Prof. Dr. Andreas Meier for his support and collaboration.

References

1. Adewole, K. S., Ajiboye, A. R., Isiaka, R. M., & Babatunde, R. S. (2012). A fuzzy logic approach for evaluation of government performance in ICT projects implementation. *Journal of Emerging Trends in Computing and Information Sciences, 2*(11), 1487–1494.
2. Almutairi, B., & Khan, A. R. (2016). Persistent architecture for optimizing web service for e-government implementation. In *2016 IEEE International Symposium on Systems Engineering (ISSE)* (pp. 1–4).
3. Andino, M., Urbina, M., & Carrasco, C. (2012). *Una Nueva Política Fiscal para el Buen Vivir: La equidad como soporte del pacto fiscal*. SRI.
4. Arora, R. K., & Gupta, M. K. (2017). e-governance using data warehousing and data mining. *International Journal of Computer Applications, 169*, 28–31.
5. Asamblea Nacional del Ecuador. (2010). *Codigo organico organizacion territorial autonomia descentralizacion*. Technical report, Asamblea Nacional del Ecuador.
6. Ayachi, R., Boukhris, I., Mellouli, S., Ben Amor, N., & Elouedi, Z. (2016). Proactive and reactive e-government services recommendation. *Universal Access in the Information Society, 15*(4), 681–697.
7. Banco Central del Ecuador. (2017). Sector Fiscal. https://www.bce.fin.ec/index.php/component/k2/item/765. Accessed September 15, 2017.
8. Benhamdi, S., Babouri, A., & Chiky, R. (2017). Personalized recommender system for e-learning environment. *Education and Information Technologies, 22*(4), 1455–1477.
9. Centro de Estudios Fiscales. (2017). Presión Fiscal. https://cef.sri.gob.ec/mod/page/view.php?id=11066. Accessed September 15, 2017.
10. Contraloria General del Estado. (2017). Consultas sobre informe aprobados. http://www.contraloria.gob.ec/WFDescarga.aspx?id=48261&tipo=inf. Accessed May 22, 2018.
11. Cornelis, C., Guo, X., Lu, J., & Zhang, G. (2005). A fuzzy relational approach to event recommendation. In *Proceedings of 2nd Indian International Conference on Artificial Intelligence (IICAI)* (Vol. 5, pp. 2231–2242).
12. Cornelis, C., Lu, J., Guo, X., & Zhang, G. (2007). One-and-only item recommendation with fuzzy logic techniques. *Information Sciences, 177*(22), 4906–4921.
13. Diario -El Comercio. (2015). La recaudación de impuestos aumentó en 5 años. http://www.elcomercio.com/actualidad/quito/recaudacion-de-impuestos-aumento-anos.html. Accessed May 22, 2018.
14. Diario -El Comercio. (2017). 42 millones más por el predial en Quito en el 2018. http://www.elcomercio.com/actualidad/impuesto-predial-recaudacion-quito-municipio.htmll. Accessed May 22, 2018.
15. Diario -El Comercio. (2018). 5 277 solicitudes se procesan para cambio de avalúo predial en Quito. http://www.elcomercio.com/actualidad/solicitudes-avaluo-predios-quito-municipio.html. Accessed May 22, 2018.
16. Diario -El Comercio. (2018). Rodas: 'He dispuesto la revisión de los avalúos de todos los predios de la ciudad que sufrieron un incremento en su valor'. http://www.elcomercio.com/actualidad/quito-revision-avaluos-reclamos-predial.html. Accessed May 22, 2018.
17. Diario -La hora. (2012). Recaudación de impuestos es mayor : Noticias de Quito : La Hora Noticias de Ecuador, sus provincias y el mundo. https://lahora.com.ec/noticia/1101442386/recaudacic3b3n-de-impuestos-es-mayor. Accessed May 22, 2018.
18. Donzé, L., & Meier, A. (2012). Applying fuzzy logic and fuzzy methods to marketing. In *Fuzzy methods for customer relationship management and marketing* (pp. 1–14). Hershey: Business Science Reference.
19. European Commission. (2015). eGovernment & Digital Public Services. https://ec.europa.eu/digital-single-market/en/policies/egovernment. Accessed September 15, 2017.
20. Felfernig, A., & Burke, R. (2008). Constraint-based recommender systems: Technologies and research issues. In *Proceedings of the 10th International Conference on Electronic Commerce, ICEC '08* (pp. 3:1–3:10). New York: ACM.

21. Gräßer, F., Malberg, H., Zaunseder, S., Beckert, S., Schmitt, J., Klinik, S. A., et al. (2016). Application of recommender system methods for therapy decision support. In *IEEE 18th International Conference on e-Health Networking, Applications and Services (Healthcom)* (pp. 1–6). Piscataway: IEEE.
22. Guo, X., & Lu, J. (2007). Intelligent e-government services with personalized recommendation techniques. *International Journal of Intelligent Systems, 22*(5), 401–417.
23. Hsu, T.-H., Chu, K.-M., & Chan, H.-C. (2000). The fuzzy clustering on market segment. In *The Ninth IEEE International Conference on Fuzzy Systems, FUZZ IEEE 2000* (Vol. 2, pp. 621–626). Piscataway: IEEE.
24. Inmon, W. H. (2002). *Building the data warehouse*. Hoboken: Wiley.
25. Jones, E. E., & Chikezi, D. (2016). Assessment of the impact of tax reforms on economic growth in Nigeria. *Journal of Accounting and Financial Management, 2*(2). ISSN 2504-8856.
26. Lu, J., Shambour, Q., Xu, Y., Lin, Q., Zhang, G. (2013). A web-based personalized business partner recommendation system using fuzzy semantic techniques. *Computational Intelligence, 29*(1), 37–69.
27. Lu, J., Wu, D., Mao, M., Wang, W., Zhang, G. (2015). Recommender system application developments: a survey. *Decision Support Systems, 74*, 12–32.
28. Meier, A., & Stormer, H. (2009). *eBusiness & eCommerce: Managing the digital value chain*. New York: Springer.
29. Molnár, B. (2017). Proposal for application of data science methods in e-government. In A. Kő & E. Francesconi (Eds.), *Electronic government and the information systems perspective* (pp. 143–157). Cham: Springer.
30. OECD. (1996). Definition of Taxes. http://www.oecd.org/daf/mai/pdf/eg2/eg2963e.pdf. Accessed September 15, 2017.
31. Paz y Miño, J. (2015). *Historia de los Impuestos en Ecuador*. Quito: SRI-PUCE.
32. Porcel, C., Ching-López, A., Tejeda-Lorente, A., Bernabé-Moreno, J., & Herrera-Viedma, E. (2018). Analysis of different proposals to improve the dissemination of information in university digital libraries. In J. Kacprzyk, E. Szmidt, S. Zadrożny, K. T. Atanassov, & M. Krawczak (Eds.), *Advances in fuzzy logic and technology 2017* (pp. 195–206). Cham: Springer.
33. Quito Informa. (2018) Sistema para revalúo de Impuesto predial está listo y operativo. http://www.quitoinforma.gob.ec/2018/02/19/sistema-para-revaluo-de-impuesto-predial-esta-listo-y-operativo/. Accessed May 22, 2018.
34. Rist, T., Seiderer, A., Hammer, S., Mayr, M., & André, E. (2015). Care - extending a digital picture frame with a recommender mode to enhance well-being of elderly people. In *2015 9th International Conference on Pervasive Computing Technologies for Healthcare (PervasiveHealth)* (pp. 112–120).
35. Schafer, J. B., Konstan, J. A., & Riedl, J. (2001). E-commerce recommendation applications. In *Applications of data mining to electronic commerce* (pp. 115–153). Berlin: Springer.
36. Servicio de Rentas Internas. (2015). Derechos del Contribuyente. http://www.sri.gob.ec/web/guest/derechos-del-contribuyente. Accessed September 15, 2017.
37. Tamayo, L. F. T. (2014). *SmartParticipation*. Berlin: Springer.
38. Terán, L., & Meier, A. (2010). A fuzzy recommender system for eElections. In *EGOVIS* (pp. 62–76). Berlin: Springer.
39. Werro, N. (2015). *Fuzzy classification of online customers*. Berlin: Springer.
40. Wu, D., Lu, J., & Zhang, G. (2015). A fuzzy tree matching-based personalized e-learning recommender system. *IEEE Transactions on Fuzzy Systems, 23*(6), 2412–2426.
41. Yager, R. R. (2003). Fuzzy logic methods in recommender systems. *Fuzzy Sets and Systems, 136*(2), 133–149.
42. Yager, R. R., & Yager, R. L. (2012). Fuzzy soft social network modeling and marketing. In *Fuzzy methods for customer relationship management and marketing: Applications and classifications* (pp. 16–40). Hershey: Business Science Reference.
43. Zadeh, L. A. (1965). Fuzzy sets. *Information and Control, 8*(3), 338–353.

44. Zadeh, L. A. (1983). A computational approach to fuzzy quantifiers in natural languages. *Computers & Mathematics with Applications, 9*(1), 149–184.
45. Zhang, S., Xi, C., Wang, Y., Zhang, W., & Chen, Y. (2013). A new method for e-government procurement using collaborative filtering and Bayesian approach. *The Scientific World Journal, 2013*, Article ID 129123, 10pp. https://doi.org/10.1155/2013/129123

Chapter 4
Fuzzy Based Investment Portfolio Management

Mayank Pandey, Vikas Singh, and Nishchal K. Verma

4.1 Introduction

In the present times, the process of investing has become complicated with lot of factors and variables to be considered before making a final decision. With the advancement in the form of digitization of economy, the opportunities and the venues available for investing have increased manifold. The age of digitization has led to generation of large amount of data in every area affected by it. This heap of data has on one hand has been a helpful resource in getting the information required. On the other hand, extracting useful information from humongous amount of raw data is indeed a challenging task [26]. Previously, investment was done based on a person or a group of persons' intuition or judgment based on the fundamental information. In the present times, there are various methods to extract trends and information from the relevant dataset. However, a prominent source of information even today is in linguistic form coming from human beings which are the experts of the financial investment field. This is the point at which the fuzzy logic can find its way into the area of investing. Today, we have vast amount of data, but not all data is precise and can be directly quantified. A lot of information is still in vague, imprecise and ambiguous form which cannot be dealt with statistical and numerical methods. Thus the use of fuzzy in this field can process this information in a form that can be computationally analyzed [4, 23, 24].

The process of making diversified investment is also popularly referred to as creating a portfolio. A portfolio consists of various investment products which include but not limited to stock equity, debt equity, financial derivatives and commodities like gold, silver, etc. The management of such portfolio is done based

M. Pandey (✉) · V. Singh · N. K. Verma
Department of Electrical Engineering, Indian Institute of Technology Kanpur, Kanpur, India
e-mail: pandeym@iitk.ac.in; vikkyk@iitk.ac.in; nishchal@iitk.ac.in

A. Meier et al. (eds.), *Applying Fuzzy Logic for the Digital Economy and Society*,
Fuzzy Management Methods, https://doi.org/10.1007/978-3-030-03368-2_4

on finding a balance between the investor's expectations of return and his appetite for the risk he/she is willing to take. The desire is always to invest such that to get maximum possible return with minimum possible risk. Such a scenario is possible only if one has maximum possible information about the stock in question. To extract a trend and prediction out of such kind, fuzzy logic and its applications can play a significant role. A lot of analysis and information is required to maintain and manage the portfolio in order to invest the appropriate amount and maintain an optimal balance of different products to generate maximum profit for the investor. As we proceed into the chapter, we will come across various fuzzy based methods for different aspects of an investment portfolio management.

4.2 State-of-the-Art Fuzzy Logic in Finance

In an investment portfolio, the stocks or financial products are picked through the information gathered from analysis of various fundamental and technical indicators. The fundamental components include overall broad performance of the company in terms of profit and growth. On the other hand, technical indicators encompass the information and pattern extracted from the time series for the factors like price and volume of the stock price. Now, fuzzy based systems are being applied to find the pattern in these time series by quantifying the experts opinions. To make the systems more robust, hybrid systems of probabilistic-fuzzy models are preferred [9].

4.2.1 Fuzzy Logic for Technical Analysis

Fuzzy based approach can be used to measure the degree of effectiveness of the technical indicator patterns. Trading strategy using these patterns was applied on buy and sell signals based on pattern recognition and statistical inference methods. A buy signal was generated when the price crossed a minimum threshold level and a sell signal was generated when the price fell from a separate threshold level. In [37], fuzzy is used to percept human cognitive ambiguity into pattern detection and analyzing. To define a pattern template, a sequence of five local extrema are used. In [13], eight pattern templates were proposed. They are Head and Shoulders, Inverse Head and Shoulders, broadening tops, broadening bottoms, triangle tops, triangle bottoms, rectangle tops and rectangle bottoms. A sequence of five consecutive local extrema E_1, E_2, E_3, E_4 and E_5 were used to define and identify these pattern templates.

From the information given in the [13], for head and shoulders pattern E_1 is a point of maxima and $E_3 > E_1$ and $E_3 > E_5$. Also, if we look at their ranges, the values of E_1 and E_5 lie within 1.5% of their combined average, which is also the case with the combination of E_2 and E_4. Similarly, if we look at the description of rectangle tops pattern then E_1 is also a point of maxima and tops and bottoms are within 0.75% of their respective averages. Also, the value of lowest top is greater than that of highest bottom. In spite of these definitions being straightforward, their crisp nature suffers from shortcoming of not including human perception involving ambiguity and reasoning.

To overcome this, the membership function for these patterns is introduced thereby using membership values. For instance, the fuzzification of Head and shoulder patterns is followed using the variable x defined as $x = \frac{E_3 - Ave_1}{Ave_1 - Ave_2}$. Here, $Ave_1 = \frac{(E_1+E_5)}{2}$, the average of the maximums and $Ave_2 = \frac{(E_2+E_4)}{2}$, average of the minimums. Making use of these values, the variable defined gives the indication of how much high "the head" is above "the shoulders" or the relative distance between the maximums or the minimums. In the similar way, the fuzzification of the other patterns can be done defining the variables based on the extrema values. In [37], applying these algorithms on a random sample of around 1400 stocks over a period of 40 years till 200 detected more than 40,000 patterns in the stock price movement. Due to introduction of linguistic variables, the patterns can now give prediction in sync with the human verbal instructions and much clarity can be there on the decision making over whether to buy, sell or hold the stocks.

4.2.2 *Rate of Return Forecasting Through Information Using Rule Extraction*

In the area of investing, the information comes from the various sources. The source for the information data comes from the investment websites and forums, analysts and experts, daily trading data and so on. The kind of these sources encompass the quantitative data, investor sentiments and expert recommendations. All of these cannot be processed through traditional statistical models due to presence of ambiguity, imprecision and linguistic variables[6]. Thus fuzzy is used in here for feature selection from the sources of information.

The recommendation of experts is selected as the experts' features. From the Internet sources, the features selected were of two types, one the number of messages towards a particular investment product. Other was the sentiments in these messages, whether positive or negative based on the linguistic input taken from these messages [2]. The direction of users' opinion is decided after performing sentiment analysis of the messages across the web sources using lexicon dictionaries [22]. Then a rule based model is adopted which is expressed in the form of

$$\text{IF} < \text{Premise} > \text{THEN} < \text{Consequent} > \tag{4.1}$$

Here $<$ Premise $>$ refers to a set of pairs containing the features and their corresponding values and $<$ Consequent $>$ refers to the label which denotes the stock returns. The return can be expressed as a function of time through the following formula $r(t) = \frac{P(t_0+t)-P(t_0)}{P(t_0)}$. Here, $P(t)$ refers to stock price at time t.

This t can be measured in seconds, hours or even days. The features were having the discrete values [6], then these values were classified into some labels with membership functions $A_1, A_2, \ldots, A_i$ which meant the status like "very small", "small", "large", etc.

This data is separated into the given number of labels or membership functions with intervals based on the mean and standard deviation values. After having fuzzy membership functions, the rule base was created. An instance of a rule for obtaining return can be like:

$$\text{IF}(X_1 \text{ is } A_1 \text{ AND } X_2 \text{ is } A_3) \text{ THEN (return is } R_2) \tag{4.2}$$

Here, X_1 and X_2 are the feature values. Once we have the rule model for a particular set of stocks, we can make the predictions for the estimated return based on the range of the feature values. For applying this algorithm, in [6] data set was taken from Shanghai stock market for a period of 1 year from 2009 to 2010. It was divided into training and test set. Apart from data, more than 85,000 recommendations were collected from around 3300 analysts from a well known Chinese financial website (http://finance.sina.com.cn/).

Also, more than 100,000 posts both positive and negative about stocks were collected from a stock message board (http://guba.eastmoney.com). The method was applied on stocks separate into having single as well as multiple information sources. These were further divided into different feature groups of small and big, young and old, dividend and no dividend, etc. Results showed that accuracy of prediction was better for multiple information sources. This highlighted the fuzzification of linguistic sources.

4.2.3 *Use of Fuzzy Time Series in Investment Analysis and Forecasting*

The fuzzy logic was introduced by Zadeh [36]. Based on that and the concept of mathematical time series, the fuzzy time series was defined. The observations or the values for these series are linguistic in nature. A fuzzy time series can be of two kinds, time-variant and time-invariant. In [27], a theoretical framework on fuzzy time series is being provided in which the historical data is linguistic in nature. If it can be understood in terms of weather, the general quantitative indicators are temperature, humidity, etc. The numerical values for say temperature can be clubbed into the fuzzy ranges of "very cold", "cold", "normal", "hot", "really hot", etc. Forecasting whether a value will fall within a range can be done with much better results.

If we look at the definition of fuzzy time series, suppose $U(t)\ \forall\ t \in [0, \infty)$ is a universe of discourse being a subset of $\Re^1$. A number of fuzzy sets $A_i(t);\ \forall\ t \in [0, \infty)$ are defined on this universe of discourse with their collection represented by

$AT(t)$. Then $AT(t)$ can be termed as a time series on the universe of discourse $U(t)$. Although the observations in a fuzzy time series are fuzzy in nature as opposed to the conventional time series, but a fuzzy time series is not fuzzy in nature. For seeing the effect of fuzzy sets on time series, suppose a fuzzy set $A_i(t) \in AT_i(t)$ is considered. If there is existence of $A_k(t-1) \in AT_k(t-1)$ and it is related to $A_i(t)$ through a fuzzy relation $R_{ik}(t, t-1)$ in the form of $A_i(t) = A_k(t-1)\, oR_{ik}(t, t-1)$. Here o is the max-min operator. In such a case, $AT(t)$ is said to be caused by $AT(t-1)$ only which is the basic principle of time series [5].

It can also be represented as $AT(t-1) \rightarrow AT(t)$. Now, we define $R(t, t-1) = \cup_{i,k} R_{ik}(t, t-1)$ as the fuzzy relation between all the sets of $t-1$ and t. On similar lines, for an integer $n > 0$ suppose there is pth type of fuzzy relation $R_a^p(t, t-n)$ defining the relation among the fuzzy sets as ;

$$A_k(t) = (A_{i_1}(t-1) \times A_{i_2}(t-2) \cdots \times A_{i_n}(t-n)) o R_a^p(t, t-m) \tag{4.3}$$

Now, the overall relation between fuzzy sets is defined as $R_a(t, t-n) = \cup_p R_a^p(t, t-n)$ being the fuzzy relation between $AT(t)$ and $AT(t-1), AT(t-2), \cdots AT(t-n)$. This fuzzy time series relation can be denoted in terms of equivalent fuzzy set elements as;

$$A_{i_1}(t-1) \cap A_{i_2}(t-2) \cap \cdots \cap A_{i_n}(t-n) \rightarrow A_k(t) \tag{4.4}$$

Equivalently, it can also be expressed in terms of $AT(t-1) \cap AT(t-2) \cap \cdots \cap AT(t-n) \rightarrow AT(t)$. On similar lines, fuzzy set elements and axiomatically fuzzy sets can also exist in fuzzy time series with union of previous sets. This can be shown in the equation as;

$$AT(t-1) \cup AT(t-2) \cup \cdots \cup AT(t-n) \rightarrow AT(t) \tag{4.5}$$

The relation between the sets is defined by;

$$AT(t) = (AT(t-1) \cup AT(t-2) \cdots \cup AT(t-n)) o R_o(t, t-m) \tag{4.6}$$

Here, $R_o(t, t-m) = \cup_p R_o^p(t, t-m)$ is fuzzy relation for union case. For union relation, it signifies that $AT(t)$ is caused by only one of $AT(t-1)$ or $AT(t-2)$ or $\cdots$ or $AT(t-n)$. Thus it is called a first order fuzzy time series model of $AT(t)$. On the other hand, the intersection or multiplication signifies that $AT(t)$ is caused by $AT(t-1), AT(t-2), \cdots AT(t-n)$ collectively. In this case, it will be called the nth order fuzzy time series model of $AT(t)$.

Now, it was mentioned that fuzzy time series is of two types, time-invariant and time-variant. This can be determined through fuzzy relation. If the fuzzy relation $R(t, t-n)$ is independent of time which means for different t_1 and t_2, $R(t_1, t_1-n) = R(t_2, t_2-n)$ then the fuzzy time series is time invariant. If that is not the case, then it will be time variant in nature. These fuzzy relations can be calculated through max-min operator method. For instance, in a first order fuzzy time series model, the relation is connected to the fuzzy set values at different instances through the equation below;

$$A_k(t) = (A_{i_1}(t-1) \cup A_{i_2}(t-2) \cup \cdots \cup A_{i_n}(t-n)) o R_o^p(t, t-n) \quad (4.7)$$

This can be represented through fuzzy IF-THEN statement as;

$$\text{IF } A_{i_1}(t-1) \text{ OR } A_{i_2}(t-2) \text{ OR } \cdots \text{OR } A_{i_n}(t-n) \text{ THEN } A_k(t). \quad (4.8)$$

From Eqs. (4.7) and (4.8), the fuzzy relation will be defined as;

$$R_o^p(t, t-n) = A_{i_1}(t-1) \times A_k(t) \cup A_{i_2}(t-2) \times A_k(t) \cup \cdots \cup A_{i_n}(t-n) \times A_k(t) \quad (4.9)$$

From this equation, we can find the relation as;

$$R_o^p(t, t-n) = \max_m \{\min_{i_m, k}(A_{i_m}(t-m) A_k(t))\} \quad (4.10)$$

This leads to $R_o(t, t-n) = \max_{(p)}\{\max_{(m)}\{\min_{(i_m,k)}(A_{i_m}(t-m)A_k(t))\}\}$.

The time series can be used for forecasting through a series of steps. Firstly, the universe of discourse for the time series involved is defined. Then the historical data is collected which is usually in the form of linguistic values. Using this data, the fuzzy sets on the universe of discourse are defined and the fuzzy relationships are set up. Then the summation of these relationships is done to create the model. Now, the input is applied to the model to calculate the output which is the forecasted value. Also, if required the output needs to be defuzzified to get a value. Thus the fuzzy time series can be useful for financial forecasting for the non linear models or when the input is available as a mixture of quantitative and linguistic values.

4.3 Portfolio Selection and Evaluation Through Fuzzy Based Systems

An Investment portfolio consists of a combination of financial products. An investor creates it either through his/her understanding or through the experts of this field. There is need for maximum possible accurate prediction to maximize profit and minimize risk. Now, the traditional mathematical models are generally linear in nature. On the other hand, the real world financial problems are mostly non linear in nature. So, traditional quantitative and statistical models often have to work with some assumptions. Also, a significant part of the input is in the form of linguistic and verbal in nature. To process that information in the form that a computer can understand, use of fuzzy systems can come to the rescue for such form of computation and analysis [21]. There are various methods in fuzzy being proposed for the same purpose for different aspects of selection and evaluation with few listed as below.

4.3.1 *Volatility Forecasting Through Self Optimal and Fuzzy C-Means Clustering Techniques*

4.3.1.1 Fuzzy C-Means Clustering

In products involving stock equities and its derivatives, creating the model for forecasting the volatility for a particular stock plays a very important role. The volatility measurement is done in two ways, daily return volatility and realized volatility. For daily return volatility, as the name suggests gives the measure in terms of daily squared returns. Whereas, the realized volatility is calculated from the summation of intra day trading high frequency returns within a day. It has been observed that the realized volatility gives better and more precise performance for the volatility models with clear depiction of every ups and downs. For clustering, in [18], a data set $X = \{x_1, x_2, \cdots, x_n\}$ is being clustered into c different subgroups in such a way that each of these groups represent a natural group. These can be arrayed as a $(c \times n)$ matrix C. A clustering algorithm C_A finds the combination of data into relevant clusters that extract the information for the best possible explanation of the structure.

Fuzzy c means (FCM) clustering is one such used fuzzy clustering model [25]. This algorithm assigns memberships to $\{x_k\}$ which are inversely proportional to their relative distance from the cluster centers c_k [8]. For instance, suppose the number of each cluster is $c = 3$. If x_k is equidistant from these three prototypes, then the membership of x_k for each of the cluster will be the same $(\frac{1}{3})$. Now coming back to volatility, realized volatility with jump models are of great importance to a risk averse investor. Now as proposed by Maciel et al. [15] the change in the asset price is expressed in terms of locally bounded variation, stochastic volatility, Standard Brownian motion and value of the discrete price movements. Thus, in realized volatility model, the cumulative daily return on the investment will be represented as;

$$r(k, 0) = \int_0^k \sigma^2(s)ds + \sum_{0 \leq s \leq k} \kappa_\sigma^2(s) \tag{4.11}$$

Here, σ refers to the overall stochastic deviation and κ amounts to the squared size of the discrete price movements in the intervals from $t = 0$ to $t = k$. The daily realized volatility can be calculated from the returns shown previously through the formula as;

$$RVOL_{k+1}(T) = \sum_{i=1}^{1/T} r_{k+i.,T}^2 \tag{4.12}$$

In this case $r_{k,T}$ is the T period return. Now, the Takagi-Sugeno (TS) fuzzy model with input dataset will be having it's rule base in the following format [32, 33].

$$R_i : \text{ IF } X \text{ is } A_i \text{ THEN } y_i = p_{i0} + p_{i1}x_1 + \cdots + p_{in}x_n \tag{4.13}$$

Here, $X = [x_1, x_2, \cdots, x_n]^T \in \Re^n$ is the input data set and R_i is the ith fuzzy rule with y_i its output. A_i is the fuzzy set for the input data for ith fuzzy rule with the membership function $\mu_{A_i}(X) : \Re^n \rightarrow [0, 1]$. There are total c clusters of data present leading to formation of c membership functions. The output generated from the TS fuzzy model is being given as;

$$y = \sum_{i=1}^{c} \left(\frac{\mu_{A_i}(X) y_i}{\sum_{j=1}^{c} \mu_{A_j}(X)} \right) \tag{4.14}$$

The matrix of cluster centers can be represented as $C_M = [c_1, c_2, \cdots, c_c]^T \in \Re^{c \times n}$. The c-means clustering algorithm generates the cluster, fuzzy partition and typicality matrices as the solution to optimize the difference between the data sets and the cluster center vectors as a part of optimizing the cost function. The solution is such that as to minimize the distance between data sets and cluster centers $D_{ik} = ||X_k - c_i||^2_{A_{ik}}$. Here, the matrix $A_{ik} = [\rho_i det(F_{ik})]^{1/n} F_{ik}^{-1}$ provides information about the cluster shape and orientation. The term ρ_i is the ith cluster volume and F_{ik} is the fuzzy dispersion matrix. The value for the fuzzy dispersion matrix can be calculated as;

$$F_{ik} = \frac{\sum_{k=1}^{m} u_{ik}^{\eta_f} (X_k - c_i)(X_k - c_i)^T}{\sum_{k=1}^{m} u_{ik}^{\eta_f}} \tag{4.15}$$

Here η_f is parameter associated with membership degrees with default value 2. Input rule base learning is done through recursive probabilistic fuzzy clustering algorithm with Mahalanobis distance D_{ik}. Now, after reading the input data X_k, it's distance from the existing clusters is calculated.

If the condition $D_{ik}^2 < \chi_{n,\beta}^2$ is satisfied then the nearest cluster is identified. Here, $\chi_{n,\beta}$ is $(1 - \beta)$th chi-squared distribution value with n degrees of freedom. Here β is the false alarm probability. Then the parameters of the closest cluster is updated while the other clusters are updated to move away their centers. If the condition is not satisfied, then a new cluster needs to be created. Then the rule consequent parameters are created using recursive least square algorithm. After that, the model output is created which in this case is the realized volatility. The error in the forecast is observed through the mean square and mean absolute error. Thus overall, the probabilistic modeling is used to form new clusters and remove old ones as per requirement. It also uses utility index to assess the relevance of the present cluster structure. Using realized volatility, forecasts are done for Value at Risk (VAR) model.

In [15], the proposed model was applied on dataset from equity market indexes of S&P500, NASDAQ, FTSE (UK) and DAX (Germany). The volatility forecast comparison assumed mean-squared forecast error (MSFE), mean absolute forecast error (MAFE) and mean percentage forecast error (MPFE). Thee given approach worked as better fit for realized volatility forecasting, especially for MPFE.

4.3.1.2 Self Optimal Clustering

Self Optimal Clustering (SOC) involves use of an optimized threshold function. It makes use of interpolation property. For the formation of cluster, the given data set X is normalized. Then a hypercube is used for bounding data points [31]. The jth instance of the data in X hyperspace is defined as

$$x^j = \{x_1^j, x_2^j, \cdots, x_D^j\} \tag{4.16}$$

Here D provides the dimensions of hyperspace. The x^j is normalized into $\bar{x^j}$ through the formula

$$\bar{x^j} = \frac{x^j - (x)_{min}}{(x)_{max} - (x)_{min}} \tag{4.17}$$

Here $(x)_{min}$ and $(x)_{max}$ are the set of minimum and maximum values in each dimension. Then the threshold value δ_n is defined for getting the neighborhood of the data point for nth cluster with β_n as its optimizing factor. It is computed as

$$\delta_n = \left(\frac{1}{2m}\sum_{i=1}^{m} \frac{min(x^j)}{\sum_{p=1}^{D} x_p^j}\right)(\beta_n) \tag{4.18}$$

Using this threshold, the potential value for each point of the cluster is calculated as

$$P_n^r = \sum_{j=1}^{m} exp\left[-\left(\frac{d^2(\bar{x^r}, \bar{x^j})}{\delta_n^2}\right)\right] \tag{4.19}$$

The cluster center is chosen from the points having highest potential value. Then the data points whose distance from the center is less than the threshold value are assigned to the concerned cluster leading to its formation. All the clusters are formed from the data set in this manner.

We analyze these clusters using various indices described below. The global silhouette value via the silhouette index is calculated for every formed cluster. Then the multiple iterations are performed through the method given in [31] till the maximum GSI value is reached for each cluster. GSI for a given cluster c_n with $n = 1, 2, 3, \cdots, M$ assign to each sample of c_n is a quality measure using $s(i)$ with

$i = 1, 2, 3, \cdots, N_n$ known as silhouette width.

$$s(i) = \frac{b(i) - a(i)}{\max(a(i), b(i))} \tag{4.20}$$

where, $a(i)$ is the average distance between the ith sample and all sample included in c_n and $b(i)$ is the minimum of the average distance between ith sample and all sample included in $c_k(k = 1, \cdots, M; k \neq M)$.

Now, the silhouette value S_n for the nth cluster is defined as;

$$S_n = \frac{1}{N_n} \sum_{i=1}^{N_n} s(i) \tag{4.21}$$

From this, we can calculate GSI index as

$$GSI = \frac{1}{M} \sum_{n=1}^{M} S_n \tag{4.22}$$

The ratio of the sum of compactness and separation of the clusters is a sum of individual cluster validity measures normalized through division by the fuzzy cardinality of each cluster and defined as;

$$PI = \sum_{n=1}^{M} \frac{\sum_{j=1}^{p} \mu_{jn}^2 ||\bar{x^j} - \bar{c_n}||^2}{N_n \sum_{k=1}^{M} ||\bar{c_k} - \bar{c_n}||^2} \tag{4.23}$$

The minimum-distance separation for partition validity is given by SI index. Its lower value signifies an improved classification.

$$SI = \frac{\sum_{n=1}^{M} \sum_{j=1}^{p} \mu_{jn}^2 ||\bar{x^j} - \bar{c_n}||^2}{p \min_{k,n} ||\bar{c_k} - \bar{c_n}||^2} \tag{4.24}$$

On these indices, self optimal clustering performs better amongst the clustering techniques and also can be used for volatility measuring and forecasting.

4.3.2 Fuzzy Decision Based Trading Transaction Costs Modeling

Apart from calculating the return on investment, estimation of costs associated with trading of various investment products like equities, financial derivatives, commodities to name a few. In [35], the transaction costs for different trading positions are considered for optimal selection through fuzzy portfolio re-balancing models. The objective is to achieve maximizing of mean and skewness with minimizing of variance and transaction costs.

Usually, the mean variance model is applied for portfolio selection [17]. In the fuzzy model, consideration of mean variance along with skewness and inclusion of short selling factor leads to better return and more efficient selection on investment portfolio [12, 21]. Now, if we take a look at multiple criteria balancing model for maximizing profit and minimizing the cost, the problem would be formed as;

Maximize $\sum_{x=1}^{n} r_x(w_x^b - w_x^s) = r^*$ (Return)

Minimize $\sum_{x=1}^{n}(w_x^b - w_x^s)\sigma_x^2 + \sum_{x=1}^{n}\sum_{y=1(y\neq x)}^{n}\sigma_{xy}(w_x^b - w_x^s)(w_y^b - w_y^s) = \sigma^*$ (Risk)

Maximize $\mathbb{E}[w^T(r-\bar{r})]^3 = k^*$ (Skewness)

Minimize $\sum_{x=1}^{n} w_x^s = w^{s,*}$ (Short-selling weight)

Minimize $\sum_{x=1}^{n}(tc_1 l_x^b + tc_2 l_x^s + tc_3 s_x^{ss} + tc_4 s_x^{rp})$

Subject to

$$\sum_{x=1}^{n}(w_x^b + kw_x^s + tc_1 l_x^b + tc_2 l_x^s + tc_3 s_x^{ss} + tc_4 s_x^{rp})$$

$$\begin{aligned} w_x^b &= w_{x,0}^b + l_x^b - l_x^s \\ w_x^s &= w_{x,0}^s + s_x^{ss} - s_x^{rp} \\ 0.05u_x &\leq w_x^b \leq 0.2u_x \\ 0.05v_x &\leq w_x^s \leq 0.2v_x \\ u_x + v_x &= y_x \end{aligned} \tag{4.25}$$

for $x = 1, 2, \cdots, n$.

Here $w = (w_1, w_2, \cdots, w_n)^T$ and $r = (r_1, r_2, \cdots, r_n)^T$ are the weights and returns on the individual financial products like stock equity, derivatives, etc. represented in vector form. Also, $w_{x,0}^b$ and $w_{x,0}^s$ are the weights of those products which were bought and short sold respectively by the investor prior to re-balancing of the portfolio. Whereas, w_x^b and w_x^s are the weights of the xth product bought and short sold respectively after adjustment. During the portfolio adjustment, l_x^b, l_x^s, s_x^{ss} and s_x^{rp} are the fraction of x financial products that should be bought, sold, short-sold and repurchased respectively by the concerned investor and $tc_1, \cdots, tc_4$ stand for their respective transaction costs.

Also, k is the short selling margin requirement at starting. u_x and v_x are the binary variables indicating xth product is chosen for long buying or short selling indicated by the value 1. The presence of y_x is to ensure that same product cannot be bought and sold at one time. Now, this multiple criteria optimization problem can be solved by fuzzy multi objective programming. In that, the multiple criteria is converted into a single criteria model having membership function value target of λ. The problem statement mentioned above is modified as follow;

Maximize λ

Subject to

$$\lambda \leq \frac{r^* - r^{AI}}{r^I - r^{AI}}, \frac{\sigma^* - \sigma^{AI}}{\sigma^I - \sigma^{AI}}, \frac{k^* - k^{AI}}{k^I - k^{AI}}, \frac{w^{s,*} - k^{s,AI}}{k^{s,I} - k^{s,AI}}, \frac{CO^* - CO^{AI}}{CO^I - CO^{AI}} \tag{4.26}$$

The CO^* in above equation stands for last minimizing statement with sum of all constraint conditions. The I and AI stand for ideal and anti-ideal value respectively of the concerned variable. Since, the mean-variance-skewness-short selling model is not precise in terms of future return prediction. So the return, risk and skewness are converted into fuzzy variables.

In [10], these variables are transformed into fuzzy triangular membership function. The membership function variable is represented as (a, b, c). The return, risk and skewness variables are derived in terms of these a, b and c points. For instance, the fuzzy return can be written as;

$$R_F = \sum_{x=1}^{n} \frac{(w_x^b - w_x^s)(a_x + 2b_x + c_x)}{4} \tag{4.27}$$

Now, instead of a fixed value for each instant, a fuzzy range is now defined. Similarly, the skewness is expressed as;

$$S_F = \frac{(\sum_{x=1}^{n}(w_x^b - w_x^s)(c_x - a_x))^2 \sum_{x=1}^{n}(w_x^b - w_x^s)(c_x + a_x - 2b_x)}{32} \tag{4.28}$$

The variance V_F can also be derived in terms of these values. Then the optimization problem would be to maximize/minimize these variables with the given constraints. Use of fuzzy through λ reduces the constraint on the task and introduction of membership function eases the window for prediction by giving range instead of a fixed value.

In [35], the proposed models used dataset of 144 individual stocks from Taiwan 50 Index and Mid-Cap 100 Index. After formation of portfolios from these stocks, Each of the portfolio is rebalanced every 20 days using data of previous 60 days. After comparison of models, it was found that presence of skewness along with short selling factor leads to maximum valuation of a portfolio.

4.3.3 *Fuzzy Based Asset Selection Through Technical Indicator*

For selection of various financial products like stock equity, commodities, etc. both fundamental and technical analysis is applied depending on the requirements.

Technical analysis is based on observation and inference from the time series movement of various variables like price, volume, etc. also known as Technical Indicators (TIs) through prediction. To include the human biases and reasoning in combination with the forecasting of TIs, a fuzzy decision process can be designed to create a Fuzzy-Probabilistic Hybrid system [9].

Now, the time for decision making is divided into intervals $t_{i+1} - t_i = wh$. Here wh is the frequency at which the investor or portfolio manager executes the decision to either buy, hold or sell. For decision making, the price indicators are opening price, closing price, highest price and lowest price in a specified time frame. However, there is always a presence of imprecision and ambiguity in this information due to presence of noise, liquidity problems, etc.

The TIs indicate the information about trend, volatility, volume and momentum for a particular stock. The types of representation of these TIs are Simple Moving Average (SMA), Exponential Moving average (EMA), Moving Average Convergence Divergence (MACD), Linear Regression Line (LRL), Rate of Change (RoC), etc. For given two TIs I_1 and I_2, their mutual information is given as;

$$MI(I_1, I_2) = H(I_1, I_2) - H(I_1) - H(I_2) \tag{4.29}$$

Here, $H(I)$ is the univariate and $H(I_1, I_2)$ is the bivariate entropy for the TIs. Lowest Value of MI is 0 when the TIs do not share any information amongst themselves. Now, there are more than one TIs. So the decision is taken by information from all of them using fusion systems. The probability fusion operators include taking decision based on the direction given by maximum indicators. It is done through various techniques. One is to take decision in favor of maximum votes, whether for buy or hold and sell or hold. The other approaches are to collectively optimize the decision area through minimizing of Bayesian risk and using all TIs to make decision from maximum likelihood.

Due to the vagueness of information available, the fuzzy fusion approach can be applied. It involves creating a possibility distribution for each product. They are obtained from quantitative as well as expert knowledge and shaped into known membership functions. Then the overall decision is taken from the fuzzy union of the values of all the fuzzy values. Alternatively, t-norm can be applied on the combination doing the product of possibilities; with co norm being sum of possibilities minus their product. For instance, when two TI sources I_1 and I_2 are given, the aggregated possibility $\pi(H)$ is being given as below;

$$\pi(H) = \max(\frac{\min(\pi_1(H), \pi_2(H))}{h_2} \min(1 - h_2, \max(\pi_1(H), \pi_2(H)))) \tag{4.30}$$

Here, $h_2 = \max_{H' \in \{H_h, H_b\}} \max(\pi_1(H'), \pi_2(H'))$, H_h is possibility decision of holding and H_b is possibility decision of buying. Sell or hold decision is taken directly using the above condition for all given TIs, while the buy or hold decision is taken through their ratio of max fuzzy values. Thus with the fusion of fuzzy-probabilistic systems with their methods combined, the better result in terms of stock selection is indicated.

In [9], the simulation for the given method was done on a collection of 49 French stocks of large cap companies with each portfolio containing either 5 or 10 stocks. The results generated showed that the possibility approach depending on a threshold confidence level being shred by all sources was the best possible approach to handle large number of sources and indicators; since some of these could be unreliable.

4.3.4 MIMO Fuzzy Modeling Based Interest Rate Forecasting

Prediction of future interest rates is a crucial part for the investment portfolio planning. Overall, it is essential for a lot of areas like Risk and Portfolio Management, Treasury Planning, Central Bank Policy, General Market Practitioners to name a few. In this aspect, traditional econometrics and statistical models have been bettered by Computational Intelligence based systems like that of Neural Networks and Fuzzy Systems. One such method of forecasting can be done through Evolving Fuzzy Systems which evolve in parameter organization and learning for fuzzy rule based models. These systems used methodologies to modify the rule system by replacing less informative rule with that of more informative in nature. The models used are MIMO evolving and extended Takagi-Sugeno (eTS and xTS) models [16].

For TS models, identification of structure is done through focal point estimation for the fuzzy rules. In eTS fuzzy model, continuous on-line training of the system takes place through recursive and non-iterative clustering method [1] . On the other hand, the xTS model clustering method works through recursive calculation of new data point potential without considering a constant cluster radius or spread. Now, as known the reasoning model for TS systems will be;

$$\begin{aligned} R_i &: \text{IF } x_1 \text{ is } A_1 \text{ AND } \cdots \text{AND } x_n \text{ is } A_n \\ \text{THEN } y_i &= \frac{\alpha_1 x_1 + \alpha_2 x_2 + \cdots + \alpha_n x_n}{\alpha_1 + \alpha_2 + \cdots + \alpha_n} \end{aligned} \tag{4.31}$$

This is for one output. Similar relation can be applied to other outputs. Here R_i stands for ith fuzzy rule. The tasks for model identification are finding out the rule base and the spread of the variable membership functions. Firstly, the clustering of the data is being done which gets collected continuously. For new data points, Cauchy function of first order is being used to calculate the potential with first point having potential of 1. If the potential of new point is greater than the old cluster center, then the new point gets replaced as the new cluster center. As the cluster centers get updated, the model parameters get updated through minimization of error between output and inputs combined with parameters. The optimized solution is obtained using recursive least squares (RLS) method.

In [16], the proposed methodologies were applied on data for US zero coupon bond monthly yields for a period of 22 years from 1987 to 2009. There were different sets of maturity periods of 1, 3, 6, 12, 24, 36, 60, 84 and 120 months. Forecasts made by eTS and xTS models outperformed the affine and statistical models for the same. These methods were also applied on Brazilian Bond data which was available daily and results were better than the US bond prediction.

4.4 Portfolio Functioning and Optimization Through Fuzzy Based Systems

After portfolio formation, it is absolutely essential to keep an eye on the prevailing conditions of market and situation of the products vis a vis the market. To manage a portfolio means to control and maintain a collection of financial products. For that matter, fuzzy based systems play an important role in quantifying the vague and ambiguous data.

In mean-variance models, the estimated return on a portfolio is found by individual expected return on each product combined with their relative weight in the portfolio. On the other hand, risk associated with the portfolio depends on the standard deviation from individual return with the respective weight. However, the accuracy of estimation is a major challenge for these type of models. Another obstacle is making use of all the available information which is shortcoming of traditional statistical models. For this aspect, fuzzy based system prove to be a major boon.

4.4.1 Neuro Fuzzy System Based Portfolio Functioning

If an investment portfolio is composed of a large number of financial products, optimizing it becomes a multidimensional complex task. A Neuro-Fuzzy based system making use of Memetic Algorithm can be useful for the managing of a financial portfolio [14]. Memetic Algorithm (MA) is used in optimization problems where it operates for heuristic search using population based approach. It functions by combining Genetic Algorithm (GA) and Local Search (LS). The GenSoFNN-Yager is a self organizing neuro-fuzzy system implementing Yager fuzzy inference system.

For training the neuro-fuzzy system, there are three steps in total to proceed. In the given system, there are total five layers [19, 20]. Apart from input and output layer, the remaining three layers consists of the fuzzy rule base. Out of these, the middle layer consists of single fuzzy rule in each node. While the preceding and proceeding layers have the antecedents and consequents of these rules. the system is trained through discrete incremental clustering algorithm. The procedure involves performing unsupervised cluster analysis on the raw data. After the fuzzy labels are formed for input and output variables, RuleMap Algorithm is applied for formation of the rule base. The learning is concluded with application of back-propagation algorithm to update network parameters.

Now, the given neuro-fuzzy system is used for forecasting the return on individual financial product based on past available information [28]. After getting the results, the Memetic Algorithm (MA) is applied to decide which product is to be entered in the portfolio based on the requirements. The methods used are local search methods of Simplex search of Nelder and Mead (NM) and Sequential Multi-Dimensional Search (SMD) with search region being local in nature. These methods are selected using random number generation.

Now, if only maximizing profit is aim, the MA-NM/SMD algorithm system may allocate 100% money in one single product. To avoid this, the constraint of limiting the investment on a single product needs to be added. Also, the problem can be set to minimize the Sharpe ratio which is given as;

$$S = \frac{\mathbb{E}[R] - R_f}{\sigma} \tag{4.32}$$

Here, $\mathbb{E}[R]$ is the expected return, R_f is the risk free rate and σ is the portfolio volatility. The optimization problem now is given as below.

Maximize $S = \frac{\sum_{x=1}^{n} w_x r_x - R_f}{[\sum_{x=1}^{n} \sum_{y=1}^{n} w_x w_y \sigma_{xy}]^{1/2}}$

Subject to

$$\sum_{x=1}^{n} w_x = 1 \tag{4.33}$$

Here, w_x is the weight assigned to asset in portfolio and r_x is the return on the asset. To minimize the risk, the highly volatile assets also are needed to be avoided by minimizing the portfolio standard deviation. For this purpose, the fitness function is also being set.

Minimize $\sigma_{po} = [\sum_{x=1}^{n} \sum_{y=1}^{n} w_x w_y \sigma_{xy}]^{1/2}$

Subject to

$$\begin{aligned} \sum_{x=1}^{n} w_x &= 1 \\ \sum_{x=1}^{n} w_x r_x &> R_f \end{aligned} \tag{4.34}$$

In [14], the proposed method used monthly returns values of Dow Jones Industrial Average (DJIA) from 1992 to 2006 for simulation purposes. The neuro fuzzy method combined with MA almost constantly generate portfolio with least possible number of stocks than the conventional methods like CAPM with lower risk.

4.4.2 *Fuzzy Decision Support System For Portfolio Functioning*

Keeping the combination of appropriate financial products for an investment portfolio is a crucial activity for an investor or a portfolio manager and is composed of a lot of decision making situations. In [11], a framework is being suggested that includes fuzzy theory into strategic portfolio selection to deal with ambiguous information which can be extended to investment portfolio management. The process involves three stages which are pre-evaluation, preference elicitation and data analysis and reporting.

The stage of pre-evaluation consists of selection of alternatives, setting targets and constraints, deciding the evaluation criteria and deciding the type of fuzzy integer linear programming model with relative significance of coefficients. The preference elicitation stage consists of defining linguistic variables and corresponding membership functions. Then it decides upon rating of the alternatives. The third stage involves usage of two algorithms. One is fuzzy weighted average for identifying relative advantage of one product over other and another one is fuzzy integer linear programming for selection of optimal combination. The optimal combination is done through calculation of weighted scores of alternative combination through making use of fuzzy weighted average. Then the score of each decision is obtained and given weights. After that an aggregate group score is obtained for obtaining a group result. Then the optimal combination is selected from these decisions using fuzzy integer linear programming.

4.4.3 *Fuzzy Logic for Trader Knowledge Representation*

In an investment portfolio, high frequency and daily trading is a crucial part of maintaining optimal combination and obtaining returns as desired by the investor. For such quick response task, efficiency of time series forecasting model is not up to the mark. A key part of this trading system is the supervisor agent which coordinates the various information, decisions of different agents of the system and then presents the final trading strategy to the HFT expert. Here agents refer to the trading agents of a-Trader system which provides support on investment decisions in FOREX Market [3]. The system has a range of decisions from $[-1, 1]$ where -1 means "recommend to sell", 0 means "recommend to hold" and 1 means "recommend to buy" [7]. The values in between signify the confidence level in the decision. In this aspect, the fuzzy based systems the range of probability along with the decisions is being given which also tells to sell or buy too fast or after a while. This type of interpretation is much closer to human thinking.

Some of the fuzzy based agents employed in the A-Trader system are Bollinger Fuzzy, Williams Fuzzy, Trend Linear-Reg-Fuzzy and Consensus Fuzzy. The Bollinger Fuzzy Agent works on the foundation of Bollinger Bands Indicator. These are the constraints representing volatility and define the range around a moving average. The size of these bands depend upon changes in the respective volatility with band widening with increase in volatility. In this agent, buy decision probability level is calculated around the upper band and sell decision around the lower band. The Williams Fuzzy Agent has Williams %R indicator which works on estimation of the momentum. In Trend Linear-Reg-Fuzzy Agent, the trend is approximated for data points with a basic linear equation $y = mx + c$. The probability level of buy or sell decision is found out based on whether the slope m changes from positive to negative or negative to positive. Finally, the Consensus Fuzzy agent takes the decision based on the decisions by other fuzzy logic agents.

4.4.4 Portfolio Optimization Through Fuzzy Asset Management

Allocation of appropriate financial products in an investment portfolio based on investor requirements is a task that requires dealing with lot of random and uncertain information. In regular mean-variance portfolio model, variance and Value-at-Risk (VaR) is used as a parameter for determining the risks and uncertainties associated with a particular combination of investments [34]. Now, to represent uncertainty fuzzy random variables can be used which can cover both randomness and fuzziness of the uncertain information. These random variables can be applied where the linguistic and subjective information has a certain amount of uncertainty and random behavior embedded in it. Now, for fuzzy random variable, a fuzzy number is given as $F_n : \Re \rightarrow [0, 1]$. Its alpha-cut is given as;

$$F_\alpha = \{n \in \Re | F_n \geq \alpha\} = [F_\alpha^-, F_\alpha^+] \ \forall \ \alpha \in (0, 1] \tag{4.35}$$

A fuzzy random variable will be a Fuzzy number map with a defined membership function and is represented as $\tilde{A} : \Omega \rightarrow N$. Here Ω is the sample space and N is the set of all fuzzy numbers. The condition for this to be random variable is that all the alpha-cuts of this number $\tilde{A}_\alpha^-$ and $\tilde{A}_\alpha^+$ belong to χ which is the family of all real valued integrable random variables.

The expectation and the risk measure for the given fuzzy random variable will also be a membership function possessing fuzzy number. For a given fuzzy random variable, their randomness is calculated through the probabilistic expectation. Also, their fuzziness is being found out through evaluation weights and λ-mean function

$[a, b] \rightarrow \lambda.a + (1 - \lambda)b$. Using these random variables, the risk can now be quantified even from the linguistic and subjective information. The variances and covariances are now estimated through the expectation of the fuzzy random variables. It leads to much better formulation of the risk profile of the portfolio.

4.4.5 Portfolio Optimization Through Combining Mean-Variance and Fuzzy Model

An investment portfolio consists of a range of financial products. Products related to stock market like share equity, financial derivatives like options and futures form a significant part of such portfolio. The major challenge which the investors face regarding their return on investment is the fluctuation in the stock market. Sometimes it is related to company's performance, sometimes it is related to emotions and sentiments while sometimes it is completely random in nature. If the reasons for price movement are not quantifiable, then the fluctuations can be unpredictable.

To avoid damage due to fluctuations, investors/managers tend to diversify into stocks from different sectors and countries. The aim is to reduce the risk to minimum possible level and get the maximum benefit out of diversification. To get such benefit, extended mean-variance model was formed out of using fuzzy approach in traditional mean-variance model [38]. In this, the return on investment is defined as a fuzzy number with a defined membership function. The different return values with course of time are defined in terms of percentile depending on its relative position. The Portfolio optimization model using Fuzzy semi-variance model can be constructed as following.

$\min \sigma_P^2 = \sum_{x=1}^{n}(r_{60,x} - r_{40,x} + \frac{1}{2}(d_{50,40} - d_{95,60}))w_x$

Subject to

$$\max R_P = \sum_{x=1}^{n}[\frac{1}{2}(r_{60,x} + r_{40,x}) + \frac{1}{4}(d_{95,60} - d_{50,40})]w_x \sum_{x=1}^{n} w_x = 1 \ \forall \ w_x \geq 0 \tag{4.36}$$

For the given problem, R_P and σ_P^2 are the fuzzy portfolio return and risk respectively. Here $r_{60,x}$ and $r_{40,x}$ are the returns at 60th and 40th percentile respectively. While $d_{95,60}$ is the spread between 95th and 60th percentile return. Similarly, $d_{50,40}$ is the spread between 50th and 40th percentile return.

Now in the conventional portfolio system using average and variance, the rate of return was assumed to have normal distribution [17]. In fuzzy semi-variance model above, this return is taken to be a fuzzy number. In extended mean variance model, the fuzzy return as well as relative risk factor between products in the portfolio in the form of co-variances are taken [29, 38]. The formulation of this model is done as below.

$\min \sigma_P^2 = \sum_{x=1}^{n} \sum_{y=1}^{n} w_x w_y Cov_{x,y}$
Subject to

$$R_P \geq \sum_{x=1}^{n} [\frac{1}{2}(r_{60,x} + r_{40,x}) + \frac{1}{4}(d_{95,60} - d_{50,40})] w_x$$
$$\sum_{x=1}^{n} w_x = 1 \ \forall \ w_x \geq 0 \tag{4.37}$$

Here $Cov_{x,y}$ is the covariance between product x and product y. All the other symbols have the same meaning as defined in the semi-variance model. Now to measure the return vs amount and type of diversification in the investment portfolio, there are different kinds of indexes which can be applied [30]. Some of those are Sharpe's, Treynors, Risk Adjusted Return and Efficient Frontier Index. The popular among them is the Efficient Frontier Index (EFI). It is a collection of return on different portfolio combinations which are feasible and risk level for maximum return with minimum risk for any rate of return. It having high value means high return for any amount of risk. The Formula for the EFI is being given as below.

$$EFI = (\sum_{x=1}^{n} \frac{R_x}{\sigma_x})(\sum_{x=1}^{n} \frac{R_x - R_{lowest}}{\sigma_x - \sigma_{lowest}}) \tag{4.38}$$

Here EFI stands for Efficient Frontier index, R_x and R_{lowest} are the instantaneous and the lowest return on portfolio combination x. Similarly, σ_x and σ_{lowest} are the instantaneous and lowest standard deviation for the portfolio combination x. The use of fuzzy modeling leads to partial solution of problem with normality present in conventional model.

In [38], the proposed methodology was used on 10 different types of portfolio used as sample. These were selected from listed stocks of companies in Bursa Malaysia. The data was for period of 11 years from 1998 to 2009 with each portfolio consisting of around 30 assets. The results showed that except in dividend based portfolio, extended mean variance model was beneficial and applicable in all remaining types of portfolio.

It was concluded that portfolio with high EFI value is more optimal for providing diversification benefit. Through simulations, it was found that 80% of extended mean variance portfolios had higher value than given traditional models for which only 10% had higher value. Thus this model was superior in various kinds of portfolio such as market size, earning, domestic and international ones.

4.5 Conclusion

As the digitization is being introduced in every aspect of economy, the working of the systems are becoming increasingly complex. The flow of information and data has increased manifold. Also, all of this information is not in the form which a computer system can understand directly. The challenge in this case is to convert this heap of information into useful and relevant data. Like other areas, field of investing and finance face same obstacle. The tolerance for the error is very less in this sector compared to other industries. This makes utilizing every piece and type of available information to get desired results very essential.

If the field of investing is to be discussed in particular, it has relied mainly on the mathematical and statistical models for a long period of time. The subjective information was taken as advisory in nature as it was not feasible to compute it in a proper way. So this type of data was not used in an efficient way. Thus the introduction of fuzzy systems have led to change in this condition. The capability of a fuzzy model to convert imprecise information to a quantified value can aid the decision making for the investor.

In an investment portfolio, there are a number of steps that lead to its formation and efficient management. While portfolio formation, there is need to find a balance between the expectations of return on investment and appetite for the risk. Use of different techniques in fuzzy system can provide solution for the common challenges in this area such as prediction of market volatility and transaction cost estimation for trading of products. A fuzzy model can compute the effect of non-financial vague factors on the price movement of portfolio products. It is also helpful in filling the gaps in information from technical indicators for facilitating the efficient trading of stocks. Use of fuzzy logic in technical analysis is used to quantify the human cognitive analysis for aiding pattern detection.

For multiple information sources, both quantitative and linguistic, using fuzzy rule extraction leads to improvement in prediction accuracy. For forecasting, use of time series is an essential tool. In such case, use of fuzzy time series for estimation of future linguistic variables using past data can lead to improvement in the prediction by utilizing information not possible by conventional methods. Using fuzzy based clustering to forecast the volatility in stock movement provides better result than conventional probabilistic models. The system of multiple Input Multiple Output (MIMO) can also be applied through fuzzy systems. The performance of extended and evolving TSK fuzzy systems for interest rate forecasting was superior than statistical systems for US Bond market.

After formation of an investment portfolio, its smooth functioning and having an optimal combination of products at every instant of time is also an essential task. There is a need to always keep a check on the movement of financial markets. Updating the portfolio based on every piece of available latest information is a task a good investor or portfolio manager is expected to perform. The use of Neuro-Fuzzy system using Memetic Algorithm generates the portfolio with the least possible stocks from the given pool with the maximum return thus keeping things simple for the investor.

There are also methods like fuzzy decision support system using fuzzy integer linear programming for getting an optimal combination of stocks in a portfolio. Also, applying extended mean variance model for portfolio optimization gives better results in terms of diversification benefits over conventional models like CAPM. Thus overall, it can be said that the use of fuzzy logic for an investment portfolio management can be a boon to its performance.

References

1. Angelov, P., Xydeas, C., & Filev, D. (2004). On-line identification of MIMO evolving Takagi-Sugeno fuzzy models. In *2004 IEEE International Conference on Fuzzy Systems (IEEE Cat. No.04CH37542)* (Vol. 1, pp. 55–60). Piscataway: IEEE.
2. Antweiler, W., & Frank, M. Z. (2004). Is all that talk just noise? The information content of internet stock message boards. *The Journal of Finance, 59*(3), 1259–1294.
3. Barbosa, R. P., & Belo, O. (2010). Multi-agent forex trading system. In *Agent and multi-agent technology for internet and enterprise systems* (pp. 91–118). Berlin: Springer.
4. Bharadhwaj, H., Singh, V., & Verma, N. K. (2019). Type-2 fuzzy TSK A type-2 fuzzy systems approach for clustering based identification of a T-S regression model. In *2017 International Conference on Computational Intelligence: Theories, Applications and Future Directions (ICCI-2017)* (pp. 359–374). Singapore: Springer.
5. Cheng, S. H., Chen, S. M., & Jian, W. S. (2016). Fuzzy time series forecasting based on fuzzy logical relationships and similarity measures. *Information Sciences, 327*, 272–287.
6. Duan, J., & Zeng, J. (2015). Forecasting stock return using multiple information sources based on rules extraction. In *2015 12th International Conference on Fuzzy Systems and Knowledge Discovery (FSKD)* (pp. 1183–1188). Piscataway: IEEE.
7. Korczak, J., Hernes, M., & Bac, M. (2015). Fuzzy logic in the multi-agent financial decision support system. In *2015 Federated Conference on Computer Science and Information Systems (FedCSIS)* (pp. 1367–1376). Piscataway: IEEE.
8. Krishnapuram, R., & Keller, J. M. (1993). A possibilistic approach to clustering. *IEEE Transactions on Fuzzy Systems, 1*(2), 98–110.
9. Le Caillec, J. M., Itani, A., Gueriot, D., & Rakotondratsimba, Y. (2017). Stock picking by probability–possibility approaches. *IEEE Transactions on Fuzzy Systems, 25*(2), 333–349.
10. Li, X., Qin, Z., & Kar, S. (2010). Mean-variance-skewness model for portfolio selection with fuzzy returns. *European Journal of Operational Research, 202*(1), 239–247.
11. Lin, C., & Hsieh, P. J. (2004). A fuzzy decision support system for strategic portfolio management. *Decision Support Systems, 38*(3), 383–398.
12. Liu, S. Y. W. S., Wang, S. Y., & Qiu, W. (2003). Mean-variance-skewness model for portfolio selection with transaction costs. *International Journal of Systems Science, 34*(4), 255–262.
13. Lo, A. W., & MacKinlay, A. C. (1988). Stock market prices do not follow random walks: Evidence from a simple specification test. *The Review of Financial Studies, 1(1)*, 41–66.
14. Lumanpauw, E., Pasquier, M., & Quek, C. (2007). MNFS-FPM: A novel memetic neuro-fuzzy system based financial portfolio management. In *2007 IEEE Congress on Evolutionary Computation* (pp. 2554–2561). Piscataway: IEEE.
15. Maciel, L., Ballini, R., & Gomide, F. (2017). Evolving possibilistic fuzzy modeling for realized volatility forecasting with jumps. *IEEE Transactions on Fuzzy Systems, 25*(2), 302–314.
16. Maciel, L., Gomide, F., & Ballini, R. (2012). MIMO evolving functional fuzzy models for interest rate forecasting. In *IEEE Conference on Computational Intelligence for Financial Engineering & Economics (CIFEr)* (pp. 1–8). Piscataway: IEEE.

17. Markowitz, H. (1952). Portfolio selection. *The Journal of Finance, 7*(1), 77–91.
18. Pal, N. R., Pal, K., Keller, J. M., & Bezdek, J. C. (2005). A possibilistic fuzzy c-means clustering algorithm. *IEEE Transactions on Fuzzy Systems, 13*(4), 517–530.
19. Rajurkar, S., & Verma, N. K. (2017). Developing deep fuzzy network with Takagi Sugeno fuzzy inference system. In *2017 IEEE International Conference on Fuzzy Systems (FUZZ-IEEE)* (pp. 1–6). Piscataway: IEEE.
20. Refenes, A. P., Burgess, A. N., & Bentz, Y. (1997). Neural networks in financial engineering: A study in methodology. *IEEE Transactions on Neural Networks, 8*(6), 1222–1267.
21. Sevakula, R. K., & Verma, N. K. (2015). Fuzzy rule reduction using sparse auto-encoders. In *IEEE International Conference on Fuzzy Systems (FUZZ-IEEE)* (pp. 1–7). Piscataway: IEEE.
22. Sevakula, R. K., & Verma, N. K. (2017). Compounding general purpose membership functions for fuzzy support vector machine under noisy environment. *IEEE Transactions on Fuzzy Systems, 25*(6), 1446–1459.
23. Singh, D. J., Agrawal, P., Verma, N. K., Ghosh, A. K., & Malagaudanavar, A. (2018). Interval type-2 TS fuzzy model for angle of attack sensor of the aircraft. *Journal of Intelligent and Fuzzy Systems, 34*(6), 3891–3901.
24. Singh, V., Dev, R., Dhar, N. K., Agrawal, P., & Verma, N. K. (2018). Adaptive type-2 fuzzy approach for filtering salt and pepper noise in grayscale images. In *IEEE Transactions on Fuzzy Systems, 26*(5), 3170–3176.
25. Singh, V., Shambhav, A., Sevakula, R. K., Verma, N. K. (2016). A new approach for splitting datasets implemented using map reduce based fuzzy c-means clustering. In *Fourth International Conference on Business Analytics and Intelligence (ICBAI)*. Bangalore: IISc.
26. Singh, V., & Verma, N. K. (2018). Deep learning architecture for high-level feature generation using stacked auto encoder for business intelligence. In *Complex Systems: Solutions and Challenges in Economics, Management and Engineering* (pp. 269–283). Cham: Springer.
27. Song, Q., & Chissom, B. S. (1993). Fuzzy time series and its models. *Fuzzy Sets and Systems, 54*(3), 269–277.
28. Tung, W. L., & Quek, C. (2002). GenSoFNN: A generic self-organizing fuzzy neural network. *IEEE Transactions on Neural Networks, 13*(5), 1075–1086.
29. Vercher, E., Bermúdez, J. D., & Segura, J. V. (2007). Fuzzy portfolio optimization under downside risk measures. *Fuzzy Sets and Systems, 158*(7), 769–782.
30. Verma, N. K. (2012). Estimation of fuzzy measures using covariance matrices in Gaussian mixtures. *Applied Computational Intelligence and Soft Computing, 2012*, 402420. https://doi.org/10.1155/2012/402420
31. Verma, N. K., & Abhishek R. (2014). Self-optimal clustering technique using optimized threshold function. *IEEE Systems Journal, 8*(4), 1213–1226.
32. Verma, N. K., & Madasu H. (2010). Additive and non-additive fuzzy hidden Markov models. *IEEE Transactions on Fuzzy Systems, 18*(1), 40–56.
33. Verma, N. K., & Madasu, H. (2010). Adaptive non-additive generalized fuzzy systems. *Applied Soft Computing, 10*(3), 820–831.
34. Yoshida, Y. (2017). Portfolios optimization with coherent risk measures in fuzzy asset management. In *2017 5th International Symposium on Computational and Business Intelligence (ISCBI)*. Piscataway: IEEE.
35. Yu, J. R., Chiou, W. J. P., Chang, W. Y., & Lee, W. Y. (2013). Modeling transaction costs and skewness in portfolio: Application of fuzzy approach. In *2013 International Conference on Fuzzy Theory and Its Applications (iFUZZY)* (pp. 397–401). Piscataway: IEEE.
36. Zadeh, L. A. (1996). Fuzzy sets. In *Fuzzy Sets, Fuzzy Logic, And Fuzzy Systems: Selected Papers* (pp. 394–432). Singapore: World Scientific.
37. Zhou, X. S., & Dong, M. (2004). Can fuzzy logic make technical analysis 20/20? *Financial Analysts Journal, 60*(4), 54–75.
38. Zulkifli, M., Daud, M., & Omar, S. (2010). Maximizing portfolio diversification benefit via extended mean-variance model. In *2010 IEEE Symposium on Industrial Electronics and Applications (ISIEA)* (pp. 675–680). Piscataway: IEEE.

Chapter 5
Z-Numbers Based TOPSIS Similarity Methodology for Company Performance Assessment in Malaysia

Ku Muhammad Naim Ku Khalif, Ahmad Syafadhli Abu Bakar, and Alexander Gegov

5.1 Introduction

In describing uncertainty, a lot of techniques have drawn the attentions of researchers and applied scientist over last decade. Decisions are made based on information given which known as data. However, information about decision is always uncertain. In real-world phenomena, the uncertain information may consist of randomness, vagueness and fuzziness. In artificial intelligence research area, the main problems that always arise are: how to reason uncertain information precisely and: how to reason using uncertain information [12]. Fuzzy set theory is certainly one of the utmost significant subfields of modern artificial intelligence. In recent years, fuzzy set theory has been adopted standard framework to deal with imprecision in data set.

Zadeh [15] introduced fuzzy set theory in representing vagueness or imprecision in a mathematical approach. In order to do so, the foremost motivation of using

K. M. N. K. Khalif (✉)
Department of Science Program (Mathematics), Faculty of Industrial Sciences and Technology, Universiti Malaysia Pahang, Pahang, Malaysia
e-mail: kunaim@ump.edu.my

A. S. A. Bakar
Mathematics Division, Centre for Foundation Studies in Science, University of Malaya, Kuala Lumpur, Malaysia

Centre of Research for Computational Sciences and Informatics in Biology, Bioindustry, Environment, Agriculture and Healthcare (CRYSTAL), University of Malaya, Kuala Lumpur, Malaysia
e-mail: ahmadsyafadhli@um.edu.my

A. Gegov
School of Computing, University of Portsmouth, Portsmouth, UK
e-mail: alexander.gegov@port.ac.uk

A. Meier et al. (eds.), *Applying Fuzzy Logic for the Digital Economy and Society*, Fuzzy Management Methods, https://doi.org/10.1007/978-3-030-03368-2_5

fuzzy set shows its ability in appropriately dealing with imprecise numerical quantities and subjective preferences of decision makers [2]. Fuzzy numbers are represented as possibility distribution where most of the real-world phenomena exist in nature are fuzzy rather than probabilistic or deterministic [19]. It was specifically designed to mathematically represent to randomness and also provide formalised tools for dealing with imprecision essential to many real problems nowadays. Technologies nowadays have been developed in fuzzy set that have potential to support all of the steps that encompass a process of model orientation and knowledge discovery. In particular, fuzzy set theory can be used in data analysis to model vague data in terms of fuzzy set.

Type-1 fuzzy and type-2 fuzzy sets are used as a unique tool to erase these imprecision and uncertainty respectively. Uncertainty is closely connected with probability, which establishes the formal framework in machine learning systems. Uncertainty and fuzziness are very prominent phenomena in science and engineering applications, where most of researchers nowadays are often used type-1 and type-2 fuzzy set in their case studies. Some of the input data sets, we cannot describe straight away or objectively because they have different interpretations and very subjective. Even, type-1 fuzzy set cannot tackle the uncertainty component completely because the degree of membership grade of type-1 fuzzy set is focusing on imprecision only. Type-2 fuzzy set is capable to deal with uncertainty or approximate reasoning. But, these two linguistic numbers do not consider the reliability part. Zadeh [16] proposed the notion of z-number, which is an ordered pair of fuzzy numbers.

The component plays the role of a fuzzy restriction and represents the information about an uncertain variable, while the component is a reliability of component and enables to represent an idea of certainty or probability [1, 7]. The idea of z-numbers is to provide a basis for computation with numbers which are not completely reliable and is more intelligent to describe the knowledge of human beings and capable to cater uncertain information.

Multi criteria decision making (MCDM) has become a study of operations research which has been widely explored by experts or practitioners [10]. It is the process of making decision in the presence of multiple criteria or objectives. Nowadays, uncertainty affects strongly the world where much of the information on which decisions are based is uncertain [4, 18]. The concept of fuzzy TOPSIS is based on the chosen alternative that should be at the shortest distance from the fuzzy positive ideal solution (FPIS) and longest distance from the fuzzy negative ideal solution (FNIS). Fuzzy TOPSIS at present offers a solution for decision makers when dealing with real world data that are usually multi criteria and involves a complex decision making process [9]. Regarding the level of interaction of with decision makers to imprecise data collection, fuzzy TOPSIS provides good agility in the decision process.

In fuzzy TOPSIS, a vertex method is applied to calculate the distance between two fuzzy ratings, which calculate the distance of each alternative from FPIS and FNIS respectively using closeness coefficient. A higher value of the closeness coefficient indicates that an alternative is closer to FPIS and farther from FNIS.

In this paper, fuzzy TOPSIS is modified to use fuzzy similarity [14] for ranking evaluation instead of using closeness coefficient. Fuzzy similarity is used to calculate the similarity between two fuzzy ratings.

In real world decision making problems, linguistic variables tend to be very complex to handle but they make more sense than classical fuzzy numbers. Rather than using classical fuzzy numbers, the linguistic scales are expressed in a more details and flexible way by z-numbers. The membership function of type-1 and type-2 fuzzy sets have no information regarding knowledge of human beings. This issue has motivated the authors to propose fuzzy similarity based fuzzy TOPSIS technique that has capability to handle knowledge of human being and uncertain information properly using z-numbers. The proposed methodology is constructed without losing the generality of the fuzzy similarity and fuzzy TOPSIS in fuzzy environment. Also, it is applied for company performance assessment using z-numbers.

The remainder of this paper is organised as follows: Sect. 5.2 introduces the concepts z-numbers. Section 5.3 views the proposed methodology of fuzzy TOPSIS Similarity using z-numbers. Section 5.4 illustrates the implementation of proposed methodology for company performance in Malaysia. Section 5.5 summarises the results and draws conclusion.

5.2 Preliminaries

5.2.1 Fuzzy Number

Definition 5.2.1 ([13]) A triangular fuzzy number is represented by the following membership function. Figure 5.1 illustrates the representation of triangular fuzzy number.

$$\mu_{\hat{A}}(x) = (a_1, a_2, a_3; 1) = \begin{cases} \frac{x-a_1}{a_2-a_1} & \text{if } a_1 \leq x \leq a_2 \\ \frac{x-a_3}{a_2-a_3} & \text{if } a_2 \leq x \leq a_3 \\ 0 & \text{otherwise} \end{cases} \quad (5.1)$$

Fig. 5.1 A triangular fuzzy number

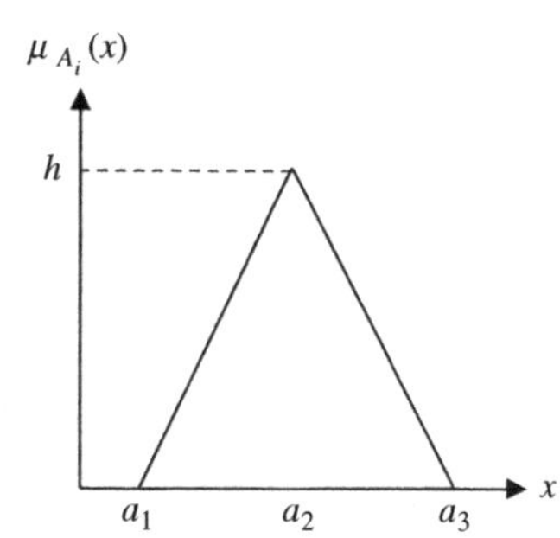

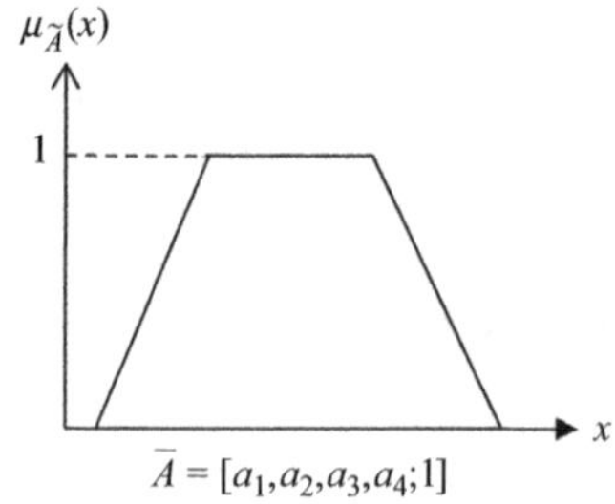

Fig. 5.2 A trapezoidal fuzzy number

Definition 5.2.2 ([13]) A trapezoidal fuzzy number is represented by the following membership function. Figure 5.2 illustrates the representation of trapezoidal fuzzy number.

$$\mu_{\hat{A}}(x) = (a_1, a_2, a_3, a_4; 1) = \begin{cases} \frac{x-a_1}{a_2-a_1} & \text{if } a_1 \le x \le a_2 \\ h & \text{if } a_2 \le x \le a_3 \\ \frac{x-a_4}{a_3-a_4} & \text{if } a_3 \le x \le a_4 \\ 0 & \text{otherwise} \end{cases} \quad (5.2)$$

5.2.2 *Z-Number*

A z-number is an ordered pair of fuzzy numbers

$$\mu_{\hat{A}}(x) = (a_1, a_2, a_3, a_4) = \begin{cases} \frac{x-a_1}{a_2-a_1} & \text{if } a_1 \le x \le a_2 \\ 1 & \text{if } a_2 \le x \le a_3 \\ \frac{a_4-x}{a_4-a_3} & \text{if } a_3 \le x \le a_4 \\ 0 & \text{otherwise} \end{cases}$$

$$\mu_{\hat{B}}(x) = (b_1, b_2, b_3, b_4) = \begin{cases} \frac{x-b_1}{b_2-b_1} & \text{if } b_1 \le x \le b_2 \\ 1 & \text{if } b_2 \le x \le b_3 \\ \frac{b_4-x}{b_4-b_3} & \text{if } b_3 \le x \le b_4 \\ 0 & \text{otherwise} \end{cases} \quad (5.3)$$

denoted as $Z\hat{A}$, $\hat{B}$. First component, $\hat{A}$ is known as restriction component whereby it is a real-valued uncertain on X while second component, $\hat{B}$ is a measure of reliability for $\hat{A}$ [16]. The illustration for z-number is depicted in Fig. 5.3 [7].

Fig. 5.3 Z-number, $Z = \tilde{A}, \tilde{B}$

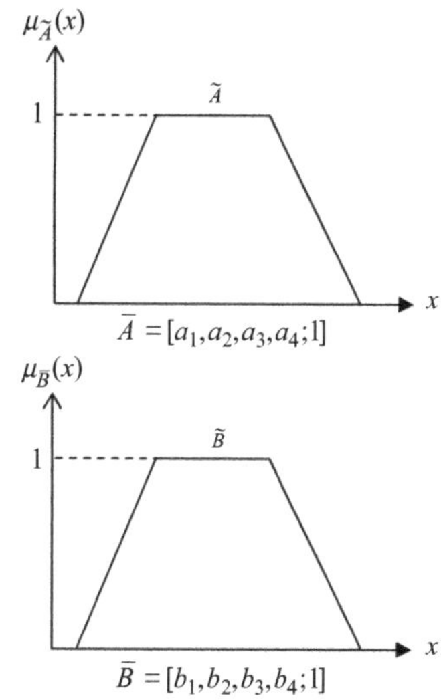

5.3 Proposed Methodology

This section focuses on the development of fuzzy TOPSIS similarity using z-numbers. The proposed methodology is extended from [5].

Step 1: Determine the Weights of Evaluation Criteria

The weighting of evaluation criteria are employed.

Step 2: Construct a Hierarchy Structure

The construction of hierarchy model shows the dependency of criteria towards alternatives that needs judgement matrix filled by decision makers about the evaluation of all criteria. Fuzzy linguistic terms are used to present the evaluation values of the alternatives preferences with respect to different criteria with degree of confidence (reliability) based on z-numbers respectively.

Step 3: Construct the Fuzzy Decision Matrix for Alternatives' Evaluation

The fuzzy decision matrix is constructed and fuzzy linguistic terms is used to evaluate the alternatives with respect to criteria.

$$\overline{DM} = \begin{array}{c} \\ A_1 \\ A_2 \\ \vdots \\ A_m \end{array} \begin{array}{c} \begin{matrix} C_1 & C_2 & \cdots & C_n \end{matrix} \\ \begin{bmatrix} \tilde{x}_{11} & \tilde{x}_{12} & \cdots & \tilde{x}_{1n} \\ \tilde{x}_{21} & \tilde{x}_{22} & \cdots & \tilde{x}_{11} \\ \vdots & \vdots & \ddots & \vdots \\ \tilde{x}_{m1} & \tilde{x}_{m2} & \cdots & \tilde{x}_{mn} \end{bmatrix} \end{array} \tag{5.4}$$

where $i = 1, 2, \ldots, m,\ j = in, 2, \ldots, n$

Step 4: Convert the Z-Numbers into Type-1 Fuzzy Numbers and Aggregate Them

All z-numbers from fuzzy decision matrices are converted into type-1 fuzzy numbers by reduction process using intuitive vectorial centroid. The intuitive vectorial centroid is an extension of the classical vectorial centroid methods for fuzzy numbers that proposed by Wen et al. [14]. Compare to other centroid methods in the literature, the way to get the centroid value is more intelligent manner, easy to compute, more balance, and consider all feasible cases of fuzzy numbers. Intuitive vectorial centroid can be computed as

$$IVC(\tilde{x}_{\tilde{A}}, \tilde{y}_{\tilde{A}}) = \left(\frac{2(a_1 + a_4) + 7(a_2 + a_3)}{18}, \frac{7h_{\tilde{A}}}{18} \right) \tag{5.5}$$

where $\tilde{x}$: the centroid point on the horizontal x-axis, $\tilde{y}$: the centroid point on the vertical y-axis, and $(\tilde{x}, \tilde{y})$: the centroid coordinate of fuzzy number $\tilde{A}$.

The reduction process of z-numbers into type-1 fuzzy sets using intuitive vectorial centroid can be computed as follows:

Assume a z-number, $Z = (\tilde{A}, \tilde{B})$, which is describe in Fig. 5.3. Let $\tilde{A} = \{\langle x, \mu_{\tilde{A}} \rangle | x \in [0, 1]\}$ and $\tilde{B} = \{\langle x, \mu_{\tilde{B}} \rangle | x \in [0, 1]\}$, $\mu_{\tilde{A}}$ and $\mu_{\tilde{B}}$ are trapezoidal membership function.

Step 1 Converting the reliability component on x-coordinate into crisp number or weight using intuitive vectorial centroid method from Eq. (5.5), $ICV_{\tilde{B}}(\tilde{x}) = \frac{2(b_1+b_4)+7(b_2+b_3)}{18} = \alpha$

Step 2 Add the weight of reliability component to the restriction component. The weighted z-number can be denoted as $\tilde{Z}^{\alpha} = \{\langle x, \mu_{\tilde{B}^{\alpha}}(x) \rangle | \mu_{\tilde{B}^{\alpha}}(x) = \alpha \mu_{\tilde{B}^{\alpha}}(x), x \in [0, 1]\}$

Theorem 5.3.1

$$E_{\tilde{A}^{\alpha}}(x) = \alpha E_{\tilde{A}^{\alpha}}(x), x \in X \tag{5.6}$$

Subject to:

$$\mu_{\tilde{A}^{\alpha}}(x) = \alpha \mu_{\tilde{A}^{\alpha}}(x), x \in X \tag{5.7}$$

Proof 2

$$E_{\tilde{A}^{\alpha}}(x) = \left[a_1, a_2, a_3, a_4; \frac{2(b_1 + b_4) + 7(b_2 + b_3)}{18} \right] = [a_1, a_2, a_3, a_4; \alpha] = \alpha E_{\tilde{A}}(x) \tag{5.8}$$

which can be denoted by Fig. 5.4 [7] below:

Step 3 Convert the irregular fuzzy number (weighted restriction) to regular fuzzy number that denoted as $\tilde{Z} = \{\langle x, \mu_{\tilde{Z}}(x) \rangle | \mu_{\tilde{Z}}(x) = \mu_{\tilde{A}}(\sqrt{\alpha} x), x \in [0, 1]\}$. In

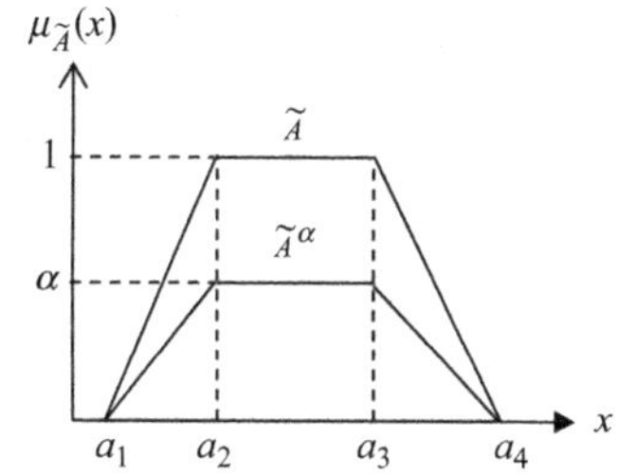

Fig. 5.4 Z-number after multiplying the reliability

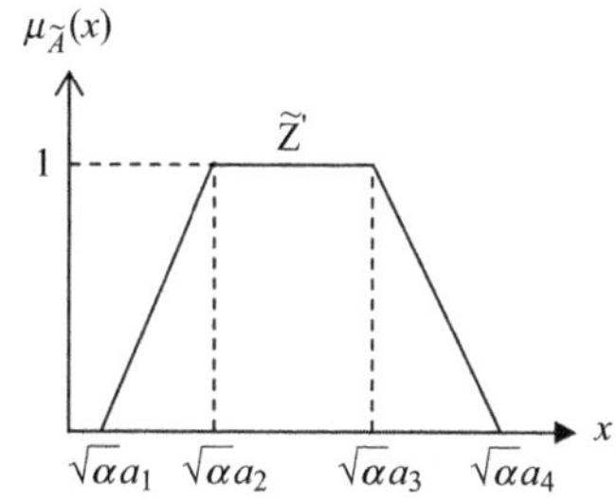

Fig. 5.5 The regular fuzzy number transformed from z-number

accordance with Theorem 5.3.3, the conclusion can be made that $\tilde{Z}$ has the same fuzzy expectation with $\tilde{Z}^{\alpha}$ where both are equal with fuzzy expectation.

Theorem 5.3.2

$$E_{\tilde{Z}'}(x) = \alpha E_{\tilde{A}}(x), x \in \sqrt{\alpha}X \tag{5.9}$$

Subject to:

$$\mu_{\tilde{Z}'}(x) = \mu_{\tilde{A}}(\sqrt{\alpha}x), x \in \sqrt{\alpha}X \tag{5.10}$$

Proof 3

$$E_{\tilde{Z}'}(x) = \left[a_1, a_2, a_3, a_4; \sqrt{\frac{2(b_1 + b_4) + 7(b_2 + b_3)}{18}}\right] = [a_1, a_2, a_3, a_4; \sqrt{\alpha}] = \sqrt{\alpha}E_{\tilde{A}}(x) \tag{5.11}$$

which can be denoted by Fig. 5.5 as follows [7]:

Theorem 5.3.3

$$E_{\tilde{Z}'}(x) = E_{\tilde{A}^{\alpha}}(x) \tag{5.12}$$

Proof 4

$$E_{\tilde{A}^{\alpha}}(x) = \alpha E_{\tilde{A}}(x) \tag{5.13}$$

$$E_{\tilde{Z}'}(x) = \alpha E_{\tilde{A}}(x) \tag{5.14}$$

$$E_{\tilde{Z}'}(x) = E_{\tilde{A}^{\alpha}}(x) \tag{5.15}$$

Then, aggregate fuzzy decision matrix from decision makers' evaluation.

$$\tilde{x}_{ij} = (\tilde{x}_{ij}^{1} \times \tilde{x}_{ij}^{2} \times \ldots \times \tilde{x}_{ij}^{n})^{1/k} \tag{5.16}$$

where $\tilde{x}_{ij}$ is the performance rating of alternatives, A_i with respect to criterion, C_j evaluated by *kth* experts and $\tilde{x}_{ij} = (a_1^k, a_2^k, a_3^k, a_4^k; h^k)$.

Step 5 Fuzzy decision matrix is weighted, averaged and normalised. Then, defuzzify the standardized generalised fuzzy numbers into coordinate form, $(\tilde{x}, \tilde{y})$.

The weighted fuzzy normalized decision matrix is denoted by $\tilde{V}$ as depicted in the next page.

$$\tilde{V} = [\tilde{v}_{ij}]_{m \otimes n}, i = 1, 2, \ldots, m; j = 1, 2, \ldots, n \tag{5.17}$$

where

$$\tilde{v}_{ij} = \tilde{x}_{ij} \times \tilde{w}_j \tag{5.18}$$

Get the average from all criteria using:

$$\tilde{v}_{ij} = \frac{1}{K}(\tilde{v}_{ij}^{1} \oplus, , , \tilde{v}_{ij}^{k} \oplus, , , \tilde{v}_{ij}^{K} \oplus) \tag{5.19}$$

Normalized each generalised trapezoidal fuzzy numbers into standardized generalised fuzzy numbers using [20]:

$$a'_{'1} = \frac{a_1 - \min(a_1, b_1)}{\max(a_4, b_4) - \max(a_1, b_1)}$$

$$a'_{'2} = \frac{a_2 - \min(a_1, b_1)}{\max(a_4, b_4) - \max(a_1, b_1)}$$

$$a'_{'3} = \frac{a_3 - \min(a_1, b_1)}{\max(a_4, b_4) - \max(a_1, b_1)}$$

$$a'_{'4} = \frac{a_4 - \min(a_1, b_1)}{\max(a_4, b_4) - \max(a_1, b_1)} \tag{5.20}$$

Step 6 Determine the fuzzy positive-ideal solution (FPIS).

Referring to normalize trapezoidal fuzzy weights, the FPIS, A^+ represents the compromise solution. The range belong to the closed interval [0,1]. The FPIS A^+ (aspiration levels) is depicted as follows:

$$A^+ = [1, 1, 1, 1; 1] \tag{5.21}$$

The FPIS, A^+ can be obtained by centroid method for (x_{A^+}, y_{A^+}).

Step 7 Calculate the similarity of each alternative from FPIS using similarity matrix.

The concept of TOPSIS method originally proposed by Hwang and Yoon [5]. Here, the authors propose fuzzy similarity to replace closeness coefficient by doing ranking evaluation. The similarity matrix is calculated based on fuzzy similarity [3]. Determine the ranking order from values of similarity measure for all alternatives using fuzzy similarity measure proposed by Hwang and Yoon [5].

$$S(\tilde{A}, \tilde{B}) = \left[1 = \left|x_{\tilde{A}*} - x_{\tilde{B}*}\right|\right] \times \left[1 - \left|h_{\tilde{A}} - h_{\tilde{B}}\right| \times \frac{\min(P(\tilde{A}), P(\tilde{B})) + \min(A(\tilde{A}), A(\tilde{B}))}{\max(P(\tilde{A}), P(\tilde{B})) + \max(A(\tilde{A}), A(\tilde{B}))}\right] \tag{5.22}$$

where, $x^*_{\tilde{A}}$ and $x^*_{\tilde{B}}$ are the horizontal centre of gravity of the generalised fuzzy numbers $\tilde{A}$ and $\tilde{B}$ that calculated using intuitive vectorial centroid in (5.5).

$P(\tilde{A})$ and $P(\tilde{B})$ are the parameter of two generalised trapezoidal fuzzy numbers $\tilde{A}$ and $\tilde{B}$, calculated as follows:

$$P(\tilde{A}) = \sqrt{(a_1 - a_2)^2 + h^2_{\tilde{A}}} + \sqrt{(a_3 - a_3)^2 + h^2_{\tilde{A}}} + (a_1 - a_2) + (a_3 - a_3) \tag{5.23}$$

$$P(\tilde{B}) = \sqrt{(b_1 - b_2)^2 + h^2_{\tilde{B}}} + \sqrt{(b_3 - b_3)^2 + h^2_{\tilde{B}}} + (b_1 - b_2) + (b_3 - b_3) \tag{5.24}$$

$A(\tilde{A})$ and $A(\tilde{B})$ are the areas of two generalised trapezoidal fuzzy numbers $\tilde{A}$ and $\tilde{B}$, calculated as follows:

$$A(\tilde{A}) = \frac{1}{2} h_{\tilde{A}}(a_3 - a_2 + a_4 - a_1) \tag{5.25}$$

$$A(\tilde{B}) = \frac{1}{2} h_{\tilde{B}}(b_3 - b_2 + b_4 - b_1) \tag{5.26}$$

The larger the value of $S(\tilde{A}, \tilde{B})$, the more the similar between two generalised fuzzy numbers $\tilde{A}$ and $\tilde{B}$.

5.4 Case Study

A case study of company performance assessment is presented here. Two experts/decision makers (DMs), DM1 and DM2 are used to evaluate 25 listed companies in Malaysia by market capital that make up the FTSE Bursa Malaysia KLCI (Last updated: 29 September 2016) [11]. Five criteria are considered to evaluate the companies which are: operation (C_1), marketing (C_2), customer (C_3), production (C_4) and, financial (C_5). Since the research problem is considered as an evaluation process, the process should involve a group of people who have expertise and knowledge in the company performance. This group is comprised of different decision makers with different level of expertise and different perceptions.

Each of decision maker has unique characteristics with regard to the evaluation process. Alongside, the decision makers usually make diverging decisions due to their different perceptions and judgements. Due to imprecise and vagueness information and the subjective nature of decision makers' judgements, which are common problems in the selection problem, uncertainty exists in the process of evaluation. In other words, the decision makers are unable to make reliable judgements regarding the evaluation procedure. Consequently, the evaluation and selection problem could be expressed as a group decision making problem under uncertain environments.

This study simplify the concept of alternatives evaluation to $\mu_{\tilde{A}} \in [0, 1]$ for fuzzy events. The values of alternatives evaluation correspond to z-numbers. The proposed Z-TOPSIS using fuzzy similarity (Z-TOPSIS-FS) is compared with Z-AHP [1] and Z-TOPSIS [8] from the literature for comparative study.

Step 1 Determine the weights of evaluation criteria. The weight of evaluation criteria are employed as same value which is 0.2 for each criterion, where the total up is 1.

Step 2 Construct a hierarchy structure. The construction of hierarchy model shows the dependency of criteria towards alternatives as presented in Fig. 5.6.

Step 3 Construct the fuzzy decision matrix for alternatives' evaluation. The fuzzy decision matrices are constructed and fuzzy linguistic terms from Tables 5.1 and 5.2 are used to evaluate the alternatives with respect to criteria. Tables 5.1 and 5.2 show the fuzzy number description used to describe the linguistic values for the restriction component and reliability component in representing z-numbers for company performance assessment by 2 decision makers. The company performance assessment are presented in Table 5.3.

Step 4 Convert the z-numbers into regular fuzzy numbers and aggregate the DMs' preferences.

The fuzzy decision matrices of DMs' preferences of z-numbers are converted and aggregated using Eqs. (5.8)–(5.15) and (5.16) respectively.

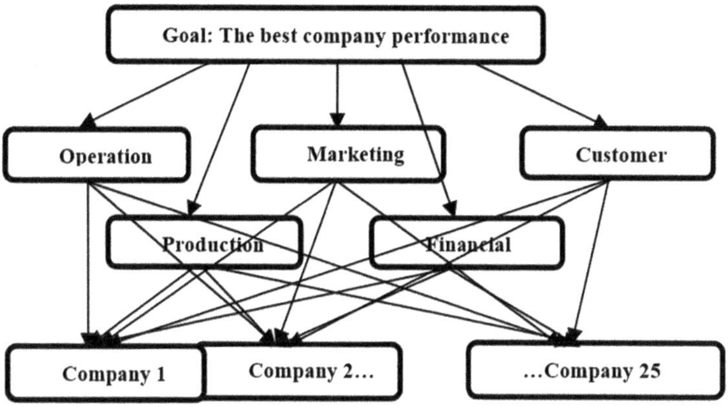

Fig. 5.6 The hierarchy of company performance assessment

Table 5.1 Linguistic terms and their corresponding generalised fuzzy numbers [17]

Linguistic terms	Generalised fuzzy numbers
Absolutely-low (AL)	(0.0, 0.0, 0.0, 0.0; 1)
Very-low (VL)	(0.0,0.0, 0.02, 0.07;1)
Low (L)	(0.04, 0.10, 0.18, 0.23; 1)
Fairly-low (FL)	(0.17, 0.22, 0.36, 0.42; 1)
Medium (M)	(0.32, 0.41, 0.58, 0.6; 1)
Fairly-high (FH)	(0.58, 0.63, 0.80, 0.86; 1)
High (H)	(0.72, 0.78, 0.92, 0.97; 1)
Very-high (VH)	(0.93, 0.98, 1.0, 1.0; 1)
Absolutely-high (AH)	(1.0, 1.0, 1.0, 1.0; 1)

Table 5.2 Reliability linguistic terms and their corresponding z-numbers [6]

Linguistic terms	Generalised fuzzy numbers
Very-low (VL)	(0,0,0,0.25;1)
Low (L)	(0,25,0.25,0.5;1)
Medium (M)	(0.25,0.5,0.5,0.75;1)
High (H)	(0.5,0.75,0.75,1;1)
Very-high (VH)	(0.75,1,1,1;1)

Table 5.3 Evaluating linguistic terms of 25 companies' performance with reliability components given by the decision makers with respect to different criteria

Company	Local performance score									
	Operation		Marketing		Customer		Production		Financial	
	DM 1	DM 2	DM 1	DM 2	DM 1	DM 2	DM 1	DM 2	DM 1	DM 2
Hap Seng Consolidated Berhad	FH (VH)	FH (VH)	M (VH)	FL (VH)	FH (VH)	FH (VH)	FH (VH)	FH (VH)	FH (VH)	H (VH)
IOI Corporation Berhad	H (VH)	H (VH)	FH (VH)	FH (VH)	FH (VH)	FH (VH)	H (VH)	H (VH)	H (VH)	FH (VH)
Sime Darby Berhad	VH (VH)	VH (VH)	H (VH)	H (VH)	H (VH)	H (VH)	VH (VH)	FH (VH)	VH (VH)	FH (VH)
Kuala Lumpur Kepong Berhad	H (VH)	FH (VH)	H (VH)	H (VH)	M (VH)	VH (VH)	M (VH)	M (VH)	M (VH)	M (VH)
Nestle (Malaysia) Berhad	M (H)	M (VH)	FH (H)	FH (VH)	M (H)	FH(VH)	FH (H)	FH (VH)	M (H)	FH (VH)
Petronas Dagangan Bhd	M (VH)	M (H)	FH (VH)	FH (H)	FH (VH)	FH (H)	H (VH)	H (H)	M (VH)	H (H)
Petronas Chemicals Group Berhad	VH (VH)	H (VH)	H (VH)	H (VH)	H (VH)	H (VH)	H (VH)	VH (VH)	H (VH)	VH (VH)
Axiata Group Berhad	H (VH)	H (VH)	FH (VH)	FH (VH)	H (VH)	H (VH)	H (VH)	H (VH)	H (VH)	H (VH)
Petronas Gas Berhad	VH (VH)	VH (VH)	H (VH)	FH (VH)	FH (VH)	VH (VH)	FH (VH)	FH (VH)	FH (VH)	H (VH)
YTL Corporation Berhad	FL (H)	FL (VH)	M (H)	FL (VH)	M (H)	M (VH)	M (H)	FL (VH)	M (H)	M (VH)
IHH Healthcare Berhad	VH (VH)	VH (VH)	VH (VH)	VH (VH)	H (VH)	H (VH)	H (VH)	H (VH)	H (VH)	FH (VH)
Hong Leong Financial Group Berhad	FH (VH)	H (H)	FH (VH)	FL (H)	M (VH)	M (H)	M (VH)	FL (H)	M (VH)	FL (H)
PPB Group Berhad	M (VH)	FH (VH)	FH (VH)	FH (VH)	M (VH)	M (VH)	FH (VH)	FH (VH)	FH (VH)	FH (VH)
CIMB Group Holdings Berhad	H (VH)	FH (VH)	FH (VH)	FH (VH)	H (VH)	H (VH)	H (VH)	H (VH)	H (VH)	H (VH)
Telekom Malaysia Berhad	M (VH)	H (VH)	H (VH)	H (VH)	FH (VH)	FH (VH)	FH (VH)	FH (VH)	M (VH)	FH (VH)
MISC Berhad	M (VH)	FH (VH)	H (VH)	H (VH)	H (VH)	FH (VH)	H (VH)	H (VH)	H (VH)	H (VH)
Public Bank Berhad	VH (H)	H (VH)	VH (H)	H (VH)	VH (H)	VH (VH)	VH (H)	VH (VH)	VH (H)	VH (VH)
Genting Malaysia Berhad	M (VH)	FH (VH)	M (VH)	FH (VH)	H (VH)	FH (VH)	H (VH)	H (VH)	H (VH)	H (VH)
Malayan Banking Berhad	VH (VH)	VH (VH)	VH (VH)	VH (VH)	H (VH)	H (VH)	H (VH)	H (VH)	H (VH	H (VH)
RHB Bank Berhad	M (VH)	M (H)	H (VH)	FL (H)	H (VH)	H (H)	FH (VH)	FH (H)	FH (VH)	FH (H)
Digi.com Berhad	H (VH)	H (VH)	H (VH)	FH (VH)	FH (VH)	FH (VH)	H (VH)	H (VH)	H (VH)	H (VH)
Hong Leong Bank Berhad	H (VH)	VH (VH)	H (VH)	H (VH)	FH (VH)	FH (VH)	FH (VH)	FH (VH)	FH (VH)	FH (VH)
Genting Berhad	FH (H)	H (VH)	M (H)	FH (VH)	H (H)	FH (VH)	H (H)	H (VH)	H (H)	VH (VH)
Tenaga Nasional Bhd	VH (VH)	H (VH)	VH (VH)	VH (VH)	VH(VH)	VH (VH)	H (VH)	VH (VH)	VH (VH)	VH (VH)
Maxis Berhad	H (VH)	H (H)	FH (VH)	FH (H)	H (VH)	H (H)	H (VH)	H (H)	H (VH)	H (H)

Step 5 Fuzzy decision matrix is weighted, averaged and normalised.

Fuzzy decision matrix is weighted, averaged and normalised using Eqs. (5.18), (5.19) and (5.20) respectively. All these results are depicted as fuzzy performance score as shown in Table 5.4.

Step 6 Determine the fuzzy positive-ideal solution (FPIS).

The FPIS, A^+ is depicted as Eq. (5.21) and can be obtained by centroid method for (x_{A^+}, y_{A^+}).

Step 7 Calculate the similarity of each alternative from FPIS using similarity matrix.

The similarity measure process is calculated using Eq. (5.22). The results of similarity measure are depicted in Table 5.4.

Table 5.4 depicts the fuzzy performance score, centroid, similarity measure and ranking result for each company performance assessment obtained by Z-TOPSIS-FS (proposed).

Table 5.5 shows that the highest ranking result for company performance assessment is Tenaga Nasional Bhd with similarity measure value 0.8265, while the lowest ranking is YTL Corporation Berhad with similarity measure value 0.1323. These two companies have the highest and lowest ranking for company performance according to the actual ranking and this is in line with the ranking obtained using Z-TOPSIS-FS (proposed), Z-AHP [1] and Z-TOPSIS [8]. These results show that the proposed technique is consistent with the actual ranking and other established techniques in the literature.

Comparison of the ranking results show that there are several but fairly minimal discrepancies in the ranking obtained by the three techniques and the actual ranking. Generally, most of the ranking results from these three techniques give quite similar. In fact, ranking for the Kuala Lumpur Kepong Berhad, YTL Corporation Berhad, Hong Leong Financial Group Berhad, Public Bank Berhad, Malayan Banking Berhad and Tenaga Nasional Bhd are same for all techniques throughout. The rest of the ranking is only slightly affected. Comparing Z-AHP [1] and Z-TOPSIS [8], both provide inconsistent ranking results for rank 9 and 11 respectively. These duplicated ranking results present the lack of ability of Z-AHP [1] and Z-TOPSIS [8] in handling linguistic assessment properly. The proposed Z-TOPSIS-FS gives prefect ordering without any duplicate ranking results. This shows that it is highly feasible to use the proposed technique in performance assessment.

Spearman's rank correlation technique is used to validate the ranking results which provides easy algebraic structure and intuitively simple interpretation. In addition, the method is less sensitive to bias due to the effect of outliers and can be used to reduce the weight of outliers (large distances get treated as a one-rank difference). In general, the coefficient of rho, (ρ) measures the strength of association between two ranked variables. The formula used to calculate Spearman's

Table 5.4 The normalised averaged weighted fuzzy performance score for each company

Company	Fuzzy performance score	Centroid (x)	Similarity measure	Ranking of performance Z-TOPSIS-FS
Hap Seng Consolidated Berhad	(0.3787,0.4519,0.6701,0.7499;1)	0.5617	0.3887	19
IOI Corporation Berhad	(0.5423,0.6155,0.8217,0.8949;1)	0.7186	0.5054	14
Sime Darby Berhad	(0.7100,0.7818,0.9122,0.9548;1)	0.8438	0.6576	6
Kuala Lumpur Kepong Berhad	(0.3787,0.4758,0.6701,0.7446;1)	0.5704	0.4006	18
Nestle (Malaysia) Berhad	(0.2765,0.3583,0.5707,0.6506;1)	0.4643	0.3218	23
Petronas Dagangan Bhd	(0.3540,0.4355,0.6369,0.7120;1)	0.5355	0.3764	20
Petronas Chemicals Group Berhad	(0.7193,0.7951,0.9335,0.9800;1)	0.8610	0.6617	5
Axiata Group Berhad	(0.5982,0.6754,0.8696,0.9388;1)	0.7716	0.5497	8
Petronas Gas Berhad	(0.6262,0.6953,0.8537,0.9069;1)	0.7727	0.5805	7
YTL Corporation Berhad	(0.0000,0.0918,0.2887,0.3710;1)	0.1892	0.1323	25
IHH Healthcare Berhad	(0.7286,0.8018,0.9282,0.9694;1)	0.8614	0.6745	4
Hong Leong Financial Group Berhad	(0.1407,0.2250,0.4234,0.5024;1)	0.3236	0.2274	24
PPB Group Berhad	(0.3454,0.4279,0.6541,0.7379;1)	0.5412	0.3688	21
CIMB Group Holdings Berhad	(0.5796,0.6554,0.8537,0.9242;1)	0.7539	0.5348	9
Telekom Malaysia Berhad	(0.4359,0.5171,0.7313,0.8098;1)	0.6239	0.4320	17
MISC Berhad	(0.5450,0.6262,0.8244,0.8962;1)	0.7242	0.5118	12
Public Bank Berhad	(0.7837,0.8488,0.9057,0.9191;1)	0.8715	0.7610	2
Genting Malaysia Berhad	(0.4732,0.5570,0.7632,0.8390;1)	0.6592	0.4602	16
Malayan Banking Berhad	(0.7472,0.8217,0.9441,0.9840;1)	0.8715	0.6915	3
RHB Bank Berhad	(0.3230,0.4193,0.6167,0.6903;1)	0.5177	0.3668	22
Digi.com Berhad	(0.5796,0.6554,0.8537,0.9242;1)	0.7539	0.5348	10
Hong Leong Bank Berhad	(0.5517,0.6222,0.8164,0.8843;1)	0.7190	0.5145	11
Genting Berhad	(0.5048,0.5781,0.7520,0.8140;1)	0.6638	0.4864	15
Tenaga Nasional Bhd	(0.8590,0.9282,0.9867,1.0000;1)	0.9512	0.8265	1
Maxis Berhad	(0.5422,0.6147,0.7971,0.8621;1)	0.7050	0.5112	13

Table 5.5 Ranking results between proposed methodology and established methods

Company	Actual rank	Z-TOPSIS-FS	Z-AHP [1]	Z-TOPSIS [8]
Hap Seng Consolidated Berhad	21	19	19	21
IOI Corporation Berhad	14	14	12	15
Sime Darby Berhad	6	6	6	7
Kuala Lumpur Kepong Berhad	18	18	18	18
Nestle (Malaysia) Berhad	23	23	23	19
Petronas Dagangan Bhd	20	20	20	22
Petronas Chemicals Group Berhad	5	5	4	3
Axiata Group Berhad	8	8	7	9
Petronas Gas Berhad	7	7	8	6
YTL Corporation Berhad	25	25	25	25
IHH Healthcare Berhad	4	4	5	5
Hong Leong Financial Group Berhad	24	24	24	24
PPB Group Berhad	21	21	21	19
CIMB Group Holdings Berhad	9	9	9	11
Telekom Malaysia Berhad	17	17	17	14
MISC Berhad	12	12	11	11
Public Bank Berhad	2	2	2	2
Genting Malaysia Berhad	16	16	16	15
Malayan Banking Berhad	3	3	3	3
RHB Bank Berhad	22	22	22	23
Digi.com Berhad	10	10	9	11
Hong Leong Bank Berhad	11	11	13	10
Genting Berhad	15	15	15	7
Tenaga Nasional Bhd	1	1	1	1
Maxis Berhad	13	13	14	17

rank is shown below.

$$\rho = 1 - \frac{6\sum \delta_i^2}{n(n^2 - 1)} \tag{5.27}$$

where

δ: the different between two ranks of each observation

n: the number of observations

The Spearman's correlation coefficient, ρ can take values between $+1$ to -1. If $\rho = 1$, indicates a perfect relationship of ranks, if $\rho = 0$, shows no relationship between ranks and $\rho = -1$, indicates a perfect negative association of ranks. The closer ρ is to zero, the weaker the relationship between the ranks. Thus based on the analysis of Spearman's rank correlation in Table 5.6, it is observed that the proposed Z-TOPSIS-FS outperforms the established Z-AHP [1] and Z-TOPSIS [8] from the literature.

Table 5.6 Ranking performance results analysis using spearman's rank correlation

Spearman's rank	Z-TOPSIS-FS	Z-AHP [1]	Z-TOPSIS [8]
ρ	0.9762	0.9750	0.9146

5.5 Conclusion

In classical fuzzy TOPSIS, evaluation depends heavily on the selection of appropriate FPIS and FNIS by using closeness coefficient. In this paper, a fuzzy similarity method is applied to company performance selection and evaluation instead of using closeness coefficient. Rationally, replacing closeness coefficient by using fuzzy similarity measure as ranking evaluation provides better judgment for representing fuzzy numbers is many aspects (centre of gravity, parameter and area). Closeness coefficient considers the distance to both FPIS (aspiration level) and FNIS (worst level), but is does not cover the more general aspects of fuzzy numbers representation, i.e. it cannot capture the vagueness of the linguistic assessment properly. The proposed Z-TOPSIS-FS provides better selection in human based decision making problems that is capable of dealing with uncertainty in human judgment. This is helpful in situations where due to lack of access to reliable information and unavailability of complete and certain data, it is hard to make right decisions. In this sense, the consideration of z-numbers in the research work provides the use of fuzzy linguistics by considering the need of human intuition in decision making problems. As a consequence, this study presents the idea in developing to design the robust and reliable methodology for selection alternatives with respect to the resources. Hence, it can be further extended by considering more complicated case studies drawn for diverse fields of human based decision making problems. To conclude, the main focus of this research study can be continued in order to make some contributions by considering real case study drawn for diverse fields crossing ecology, health, genetics, finance and so forth.

References

1. Azadeh, A., Saberi, M., Atashbar, N. Z., Chang, E., & Pazhoheshfar, P. (2013). Z-AHP: A z-number extension of fuzzy analytical hierarchy process. In *2013 7th IEEE International Conference on Digital Ecosystems and Technologies (DEST)* (pp. 141–147). Menlo Park: IEEE.
2. Deng, H. (2013). Comparing and ranking fuzzy numbers using ideal solutions. *Journal of Applied Mathematical Modelling, 38*(5–6), 1638–1646.
3. Deng, Y., Kang, B. Y., Zhang, Y. J., Deng, X. Y., & Zhang, H. X. (2011). A modified similarity measure of generalized fuzzy numbers. *Procedia Engineering, 15*, 2773–2777.
4. Gegov, A., Arabikhan, F., Sanders, D., Vatchova, B., & Vasileva, T. (2017). Fuzzy networks with feedback rule bases for complex systems modelling. *International Journal of Knowledge-Based and Intelligent Engineering Systems, 21*(4), 211–225.
5. Hwang, C.-L., & Yoon, K. (1981). *Multiple attribute decision making*. New York: Springer.

6. Kang, B., Wei, D., Li, Y., & Deng, Y. (2012). Decision making using z-numbers under uncertain environment. *The Journal of Computer Information Systems, 8*(7), 2807–2814.
7. Kang, B., Wei, D., Li, Y., & Deng, Y. (2012). A method of converting z-number to classical fuzzy number. *International Journal of Information and Computing Science, 9*(3), 703–709.
8. Krohling, R. A., Santos, A. D., Pacheco, A. G., Krohling, R. A., & Santo, E. (2016). TODIM and TOPSIS with z-numbers. Preprint. arXiv:1609.05705.
9. Madi, E. N., Tap, A. O. M., & Osman, A. (2011). Fuzzy TOPSIS method in the selection of investment boards by incorporating operational risks. In *Proceedings of the World Congress on Engineering*, London (Vol. I, pp. 6–10).
10. Mardani, A., Jusoh, A., & Zavadskas, E. K. (2015). Fuzzy multiple criteria decision-making techniques and applications—Two decades review from 1994 to 2014. *Expert Systems with Applications, 42*(8), 4126–4148.
11. MyPF.my. (2016). Bursa Malaysia Top 30. http://mypf.my/investing/equities/bursa30/
12. Tang, Y., Pan, W., Li, H., & Xu, Y. (2002). Fuzzy Naïve Bayes classifier base on fuzzy clustering. In *IEEE International Conference on Systems, Man and Cybernetics* (Vol. 5, 6 pp.). Yasmine Hammamet: IEEE Press.
13. Van Laarhoven, P. J. M., & Pedrycz, W. (1983). A fuzzy extension of Saaty's priority theory. *Fuzzy Sets and Systems, 11*(1–3), 199–227.
14. Wen, J., Fan, X., Duanmu, D., & Yong, D. (2011). A modified similarity measure of generalized fuzzy numbers. *Procedia Engineering, 15*, 2773–2777.
15. Zadeh, L. A. (1965). Fuzzy sets. *Information and Control, 8*(3), 338–353.
16. Zadeh, L. A. (2011). A note on z-numbers. *Information Sciences, 181*(14), 2923–2932.
17. Zheng, G., Zhu, N., Tian, Z., Chen, Y., & Sun, B. (2012). Application of a trapezoidal fuzzy AHP method for work safety evaluation and early warning rating of hot and humid environments. *Safety Science, 50*(2), 228–239.
18. Zhi-quan, X. (2014). Application of z-numbers in multi-criteria decision making. In *Proceedings 2014 International Conference on Informative and Cybernetics for Computational Social Systems (ICCSS)* (pp. 91–95). Qingdao: IEEE.
19. Zimmerman, H. J. (2000). An application – Oriented view of modelling uncertainty. *European Journal of Operational Research, 122*(2), 190–198.
20. Zuo, X., Wang, L., & Yue, Y. (2013). A new similarity measure of generalized trapezoidal fuzzy numbers and its application on rotor fault diagnosis. *Mathematical Problems in Engineering, 2013*, 824706.

Chapter 6
Modeling Human Perceptions in e-Commerce Applications: A Case Study on Business-to-Consumers Websites in the Textile and Fashion Sector

Adrian Castro-Lopez and Jose M. Alonso

6.1 Introduction

The fashion market is global and presents a complex structure, which operates in many different levels to reach all kinds of public, from those who love fashion to those who believe that buying clothes is a daily necessity. Fashion is a global industry with a market value of around $1.7 billion [17].

Besides, the textile and fashion sector via Internet is placed among those with higher importance revenue figures in the worldwide on-line market [22]. The main sectors of activity in Spain are detailed in [10]: Travel agents and tour operators (14.4%), Flights (11.9%) and Clothing (5.4%). Moreover, the remarkable growth rates in the clothing sector in recent years have led fashion companies to use the Business-to-Consumers (B2C) on-line channel as a mean for promoting and selling.

The success or failure of different B2C websites highly depends on the e-service quality perceived by consumers. The e-service quality can be defined as [16]:

> The extent to which a website provides effective and efficient results in regard to the information search process, to the purchase and delivery of products and services, and even to the client enjoyment and emotional experience.

In this regard, there are several models of e-service quality (e.g., ESQ and Customer Experience [20] or New PeSQ [19]). To sum up with, previous studies

---

A. Castro-Lopez
Business Administration Department, Faculty of Economic and Business, University of Oviedo, Oviedo, Spain
e-mail: castroadrian@uniovi.es

J. M. Alonso (✉)
Centro Singular de Investigación en Tecnoloxías da Información (CiTIUS), Universidade de Santiago de Compostela, Santiago de Compostela, Spain
e-mail: josemaria.alonso.moral@usc.es

A. Meier et al. (eds.), *Applying Fuzzy Logic for the Digital Economy and Society*, Fuzzy Management Methods, https://doi.org/10.1007/978-3-030-03368-2_6

have already identified different latent dimensions in e-service quality. In addition, they suggest that it is necessary to analyze hedonic and utilitarian dimensions of e-service quality. Utilitarian quality is defined as the value derived from completing objectives, from finding information, and/or from buying. Hedonic quality is defined as the value derived from enjoying the search for information and/or for purchasing. Moreover, many e-service quality models rely on inquiries to consumers about their perception on both hedonic and utilitarian dimensions. However, consumer perceptions on qualitative issues are likely to suffer from high levels of uncertainty and vagueness. Therefore, it is required to find a suitable methodology to deal with the uncertainty inherent to consumer perceptions in e-service quality assessment and modeling.

Fuzzy Logic provides a framework to the Computational Theory of Perceptions (CTP) [24] which is acknowledged for its well-known ability for approximate reasoning and linguistic concept modeling; mainly due to its semantic expressivity close to natural language. Fuzzy sets and systems are able to mathematically formalize, in an approximate but even precise way, uncertainty and vague concepts (like hedonic and utilitarian dimensions of e-service quality). In addition, interpretability of fuzzy sets and systems [2], due to its human-centric character, plays a key role in system modeling and it becomes essential in applications with high human interaction like sensory evaluation [25]. Moreover, a recent survey on the eXplainable Artificial Intelligence (XAI) research field [3] has shown the relevance of interpretable fuzzy systems in the quest for XAI systems. Notice that, the recent success of many AI applications into real-world usage has triggered some critical voices regarding ethical and legal issues. Moreover, the new European General Data Protection Regulation (GDPR[1]), approved by the European Parliament and to take effect in May 2018, refers to the "right to explanation" to European citizens. This new regulation makes even more appealing the design of XAI systems in general, and the modeling of interpretable fuzzy systems in particular, as a way to pave the way towards XAI.

The purpose of this study is to expand and further explore the knowledge on e-service quality. We combine marketing methods (qualitative and quantitative methods) and CTP (Fuzzy Logic) for the assessment and modeling of e-service quality. As a result, we get a more dynamic evaluation, enhancing adaptability to changing needs of consumer perceptions. Accordingly, business managers can redirect the investment strategies and focus on what is actually valued by consumers. Thus, in this paper we contribute to the field of analysis on e-service quality as follows:

- Data acquisition is addressed in terms of collecting consumer perceptions (regarding hedonic and utilitarian dimensions) through fuzzy rating scale-based questionnaires [18].

[1]http://eur-lex.europa.eu/legal-content/en/TXT/?uri=CELEX%3A32016R0679.

- Collected data are processed under the fuzzy logic formalism provided by CTP. Thus, we deal efficiently with uncertainty and vagueness all along the processing chain, including aggregation and fusion in the search of consensus agreement among groups of consumers. Notice that we consider the use of fuzzy multi-criteria decision-making tools [21].
- The relation between the main dimensions of e-service quality is modeled by means of a set of linguistic variables and fuzzy IF-THEN rules. The result is an interpretable fuzzy system which combines both expert knowledge and knowledge automatically extracted from data [4].
- The proposal is validated with a study of the main Business-to-Consumers websites in the Spanish textile and fashion sector.

The rest of the paper is structured as follows. Section 6.2 introduces the materials and methods applied to carry out this study. Then, Sect. 6.3 presents and discusses the main reported results. Finally, main conclusions and future perspectives are sketched in Sect. 6.4.

6.2 Materials and Methods

6.2.1 Survey Methods

Survey methods have been worldwide applied to collect opinions from consumers. Surveys supported by Likert scales [14] are likely to be the most usual ones, mainly because of their simplicity. Respondents (also called assessors in the field of sensory sciences) are usually asked to choose an answer among a small set of options (commonly expressed by ordered linguistic terms). This fact implies a lack of flexibility that is argued as the main disadvantage of this kind of surveys [6]. Moreover, the goodness of drawn conclusions strongly depends on how carefully surveys were designed in order to avoid bias and minimize ambiguity, imprecision and uncertainty in the given questionnaires. Of course, understanding properly the meaning of the involved linguistic terms depends on the context and background of each respondent. In addition, human perceptions and opinions are always subjective and it is not feasible to check how truthful respondents are. From a psychometric point of view, fuzzy rating scales make easier the assessment of the diversity, subjectivity, imprecision and uncertainty which are inherent to human perceptions [11]. Surveys supported by fuzzy rating scale-based questionnaires are especially helpful in practical applications. For example, Gil et al. applied them to teaching evaluation [12]. Moreover, Quirós et al. proved their utility in relation with the customized packaging design of gin bottles [18].

6.2.2 Quantitative and Qualitative Analysis of Human Perceptions

In a previous study [18], we proposed a new methodology for descriptive and comparative analysis of human perceptions expressed through fuzzy questionnaires. The treatment of collected data requires the adaptation of statistical techniques to the fuzzy case. It is worthy to note that SMIRE[2] researchers have actively developed statistical tools around the concept of fuzzy rating scale [11, 12]. Moreover, they have provided the research community with the free software *R* package called SAFD.[3]

Both the design of a specific fuzzy questionnaire and the analysis of collected data are made with the Quale software [5]. Quale implements the methodology described in [18] and calls to SAFD for dealing with fuzzy statistics. Moreover, it produces as result a survey report made up of a set of graphs and texts in a user-friendly style which can be customized in accordance with the reader background and preferences. Firstly, sensory data acquired through fuzzy rating scale-based questionnaires are formalized under fuzzy logic formalism. The three values that characterize each given evaluation are translated into a triangular fuzzy set $A = (a; b; c; h)$, where b represents the modal point (upper value of the fuzzy triangle), a and c determine the support (lower confidence interval), and h is the height of the triangle (by default it takes value 1). Let X be a non-empty set. Being $FS(X)$ the set of all fuzzy sets in X, $A_i = (a_i, b_i, c_i) \in FS(X)$ corresponds to the evaluation provided by assessor i. Once a set of evaluations have been collected regarding a specific sample, then they are aggregated by the sample Aumann-type mean:

$$\frac{1}{n}\sum_{i=1}^{n} A_i = \left(\frac{1}{n}\sum_{i=1}^{n} a_i, \frac{1}{n}\sum_{i=1}^{n} b_i, \frac{1}{n}\sum_{i=1}^{n} c_i\right) \tag{6.1}$$

We group those points in the scale with the greatest aggregated values until a fixed threshold of the total is reached. Then, we build the intervals that best shape the set of points. In case two or more intervals are close enough, they are fused into a single interval. Later, we compute the center of gravity (COG) of the most representative interval (that one covering most evaluations). Given a triangular fuzzy set $A \in FS(X)$, COG is calculated as follows:

$$COG(A) = \min\{y \in [a, c] | \int_a^y \mu_A(x)\mathrm{d}x \geq 0.5\} \tag{6.2}$$

[2]SMIRE stands for Statistical Methods with Imprecise Random Elements. This is the name of the Statistics and Fuzzy Logic research group in the University of Oviedo (Spain).

[3]SAFD stands for Statistical Analysis of Fuzzy Data. This *R* package is available at https://cran.r-project.org/web/packages/SAFD/index.html [Accessed on May 2018].

where $\mu_A(x)$ measures the membership degree of x to A. $COG(A)$ represents the aggregated score associated to the sample and attribute under study. In addition, the number of evaluations characterized by a fuzzy set A is given by:

$$p_A = \sum_{i=1}^{m} S(A_i, A) \tag{6.3}$$

where $S(A_i, A)$ measures the degree up to which A_i is a subset of A:

$$S(A_i, A) = 1 - \frac{\sum_{x \in X} \max(0, \mu_{A_i}(x) - \mu_A(x))}{\sum_{x \in X} \mu_{A_i}(x)} \tag{6.4}$$

The samples under study are ranked with respect to their related scores. Those samples without faithful scores are set "in quarantine" and separated from the rest. We consider three situations which denote a lack of consensus:

- The main interval is too narrow. Thus, it does not characterize a big enough number of assessors.
- The main interval is too wide.
- There exists a second interval which becomes comparable to the main interval in terms of associated evaluations.

The interested reader is kindly referred to [18] for a deeper explanation of the Quale methodology that we have only sketched above for the sake of brevity.

6.2.3 Multi-Criteria Decision-Making Tools

There are different tools for the evaluation of a group of alternatives as a function of a finite number of criteria given by a decision maker or a group of them. Some of the basic methods are [13]: Weighted Sum Model (WSM), Weighted Product Model (WPM), Compromise Programming (CP), Analytical Hierarchy Process (AHP), Elimination and Choice Expressing Reality method (ELECTRE), Technique for Order of Preference by Similarity to Ideal Solution (TOPSIS), Preference Ranking Organization Method for Enrichment Evaluation (PROMETHEE) and Serbian Multi-criteria Optimization and Compromise Solution (VIKOR).

We will focus on the TOPSIS method and its fuzzy extension F-TOPSIS [23]. This is a suitable tool to handle properly the uncertainty that is intrinsic to the opinions in a decision-making process. F-TOPSIS has been successfully used in several applications (e.g., supply chain management [9] or shopping website evaluation [21]). The F-TOPSIS method is summarized in the next three steps:

- **Step 1.** Determination of the fuzzy decision matrix: Defining the n fuzzy evaluation criteria ($C_1,\ldots,C_j,\ldots,C_n$) for all m alternatives ($A_1,\ldots,A_i,\ldots,A_m$); and building the $m \times n$ matrix:

$$\left[\widetilde{D_x}\right] = \begin{pmatrix} \widetilde{x}_{11} & \cdots & \widetilde{x}_{1j} & \cdots & \widetilde{x}_{1n} \\ \vdots & \ddots & \vdots & \ddots & \vdots \\ \widetilde{x}_{i1} & \cdots & \widetilde{x}_{ij} & \cdots & \widetilde{x}_{in} \\ \vdots & \ddots & \vdots & \ddots & \vdots \\ \widetilde{x}_{m1} & \cdots & \widetilde{x}_{mj} & \cdots & \widetilde{x}_{mn} \end{pmatrix} \tag{6.5}$$

where $\widetilde{x}_{ij} = \left(\widetilde{x}_{ij}^1, \widetilde{x}_{ij}^2, \widetilde{x}_{ij}^3\right)$ is a triangular fuzzy number which corresponds to alternative A_i and criterion C_j, with $i \in [1, m]$ and $j \in [1, n]$.
- **Step 2.** Construction of the normalized and weighted decision matrix:

$$\widetilde{x}_{ij}^* = \left(\widetilde{x}_{ij}^{1*}, \widetilde{x}_{ij}^{2*}, \widetilde{x}_{ij}^{3*}\right); \widetilde{x}_{ij}^{k*} = \frac{\widetilde{x}_{ij}^k}{\max_j \left(\widetilde{x}_{ij}^3\right)}; \forall k \in [1, 3] \tag{6.6}$$

$$\begin{gathered} \widetilde{v}_{ij} = \widetilde{W}_j \times \widetilde{x}_{ij}^*; \widetilde{v}_{ij}^k = w_j^k \times \widetilde{x}_{ij}^{k*}; \widetilde{W}_j = \left(w_j^1, w_j^2, w_j^3\right) \\ \forall k \in [1, 3]; \forall i \in [1, m]; \forall j \in [1, n] \end{gathered} \tag{6.7}$$

- **Step 3.** Closeness coefficients for each alternative and ranking.
  - Determination of the Ideal Positive Fuzzy Solution ($FPIS^+$) and the Ideal Negative Fuzzy Solution ($FNIS^-$):

$$FPIS^+ = \left\{\widetilde{v}_1^+, \ldots, \widetilde{v}_j^+, \ldots, \widetilde{v}_n^+\right\}; \widetilde{v}_j^+ = (1, 1, 1); \forall j \in [1, n] \tag{6.8}$$

$$FNIS^- = \left\{\widetilde{v}_1^-, \ldots, \widetilde{v}_j^-, \ldots, \widetilde{v}_n^-\right\}; \widetilde{v}_j^- = (0, 0, 0); \forall j \in [1, n] \tag{6.9}$$

  - Computing the distance between weighted criteria and the closeness coefficient for each alternative:

$$d_i^+ = \sum_{i=1}^{n} d(\widetilde{v}_{ij}, \widetilde{v}_j^+); \forall i \in [1, m]; \forall j \in [1, n] \tag{6.10}$$

$$d_i^- = \sum_{i=1}^{n} d(\widetilde{v}_{ij}, \widetilde{v}_j^-); \forall i \in [1, m]; \forall j \in [1, n] \tag{6.11}$$

$$CC_i = \frac{d_i^-}{d_i^- + d_i^+}; \forall i \in [1, m] \tag{6.12}$$

The final ranking of alternatives is established in accordance with (6.12). The interested reader is kindly referred to [23] for a deeper explanation about F-TOPSIS.

6.2.4 Interpretable Fuzzy Modeling

Fuzzy techniques are ready to deal properly with imprecision and uncertainty in the identification and modeling of systems [7]. Namely, the Highly Interpretable Linguistic Knowledge (HILK) methodology [4] is aimed at designing fuzzy models by combining expert knowledge (derived by human knowledge-elicitation tasks such as interviews, surveys, and so on) and knowledge automatically extracted from data (derived by data-mining tasks). This methodology is implemented in the free software GUAJE[4] [15] which makes intuitive the generation of understandable and accurate fuzzy models.

A fuzzy model is made up of two main components: the knowledge base (KB) and the inference engine. On the one hand, the KB comprises a set of linguistic variables and rules (which combine expert and induced knowledge). Notice that knowledge representation tasks are carried out off-line. On the other hand, the inference engine is in charge of exploiting the model on-line.

Regarding the construction of the KB, a panel of experts is asked to define relevant variables and rules. In addition, we can apply data mining tools provided by GUAJE because the key issue in HILK is the careful combination of expert and induced knowledge. The entire modeling process comprises three steps:

- **Fuzzy partition design.** The goal is to define the most influential variables, according to both expert knowledge and knowledge extracted from data. On the one hand, experts provide complete or partial information about the identified variables. On the other hand, several algorithms can be used to create fuzzy partitions form data. The result is the definition of a common universe for each variable according to both expert knowledge and data distribution. Notice that linguistic constraints (distinguishability, normalization, coverage, overlapping, etc.) have to be superimposed to the fuzzy partition definition in order to ensure interpretability [2]. Thus, we recommend the use strong fuzzy partitions which satisfy all previous interpretability constraints and are defined as follows:

$$\sum_{i=1}^{M} \mu_{A_i}(x) = 1, \ \forall x \in U \tag{6.13}$$

  where U=$[U_l,U_u]$ is the universe of discourse, M is the number of linguistic terms, and $\mu_{A_i}(x)$ is the membership degree of x to the A_i fuzzy set.

[4]http://sourceforge.net/projects/guajefuzzy/.

- **Rule base definition and integration.** Experts are invited to describe the system behavior through linguistic rules (*Expert Rules*). In addition, rules are induced from data (*Induced Rules*). Both types of rules use the same linguistic terms defined by the same fuzzy sets. Rule format is as follows:

$$\textbf{If} \quad \overbrace{\underbrace{X_a \text{ is } A_a^i}_{Partial\ Premise\ P_a} \quad \textbf{AND} \ldots \textbf{AND} \quad \underbrace{X_z \text{ is } A_z^j}_{Partial\ Premise\ P_z}}^{} \quad \textbf{Then} \quad \underbrace{Y \text{ is } C^n}_{Conclusion}$$

$$\underbrace{\hspace{20em}}_{Premise}$$

- **KB improvement.** The goal of this step is to enhance the KB interpretability-accuracy trade-off. First, the KB quality is assessed according to both accuracy and interpretability. Second, a simplification procedure is run to increase interpretability without penalizing either consistency or accuracy. Third, an optimization process is applied to get better accuracy while keeping interpretability.

The interested reader is referred to [4] for more details about the HILK methodology. In addition, a thorough review on fuzzy system software is given in [1].

6.3 Results

This section goes in depth with the results coming out from applying the materials and methods previously introduced to a use case regarding B2C websites in the textile and fashion sector. For the sake of readability, the section is split into two additional ones. We start with presenting and discussing results related to e-service quality analysis. Then, we focus on results related to e-service quality modeling.

6.3.1 e-Service Quality Analysis

In a preliminary study [8], we carried out a classical Likert-based survey on the Spanish textile and fashion sector. We collected opinions of a sample of 405 habitual consumers from sales platforms. The survey was disseminated through social networks (Facebook, LinkedIn and Twitter), by email and through personal interviews. The sampling error was ±2.42% with a trust level of 95% (p = q = 0.5). The sample distribution was done by levels of age (21% between 18 and 24 years, 49% between 25 and 34 years, 19% between 35 and 44 years, 11% over 45 years) and gender (60% women, 40% men).

The questionnaire was made for two groups of consumers: (1) those consumers who only search for information (40%), and those ones who search for information

Table 6.1 Ranking of B2C websites provided by F-TOPSIS

	eBay		Zara		Privalia		Buy Vip		Vente Privee		Asos		El Corte Inglés	
	d^+	d^-	d^+	d^-	d^+	d^-	d^+	d^-	d^+	d^-	d^+	d^-	d^+	d^-
C_1	0.42	0.61	0.35	0.69	0.36	0.68	0.37	0.67	0.35	0.69	0.35	0.69	0.44	0.59
C_2	0.38	0.65	0.32	0.72	0.33	0.71	0.34	0.70	0.31	0.74	0.31	0.74	0.36	0.68
C_3	0.43	0.60	0.31	0.73	0.34	0.70	0.35	0.69	0.35	0.69	0.35	0.68	0.32	0.72
C_4	0.38	0.66	0.40	0.63	0.36	0.68	0.37	0.66	0.35	0.69	0.35	0.69	0.43	0.60
C_5	0.64	0.38	0.62	0.40	0.68	0.34	0.66	0.36	0.78	0.24	0.60	0.42	0.69	0.33
C_6	0.39	0.65	0.37	0.67	0.37	0.66	0.37	0.66	0.43	0.60	0.30	0.74	0.35	0.68
C_7	0.37	0.66	0.37	0.66	0.36	0.67	0.34	0.69	0.45	0.57	0.34	0.70	0.38	0.66
C_8	0.35	0.68	0.45	0.57	0.43	0.59	0.45	0.58	0.59	0.43	0.46	0.57	0.42	0.61
C_9	0.46	0.57	0.38	0.65	0.46	0.57	0.46	0.57	0.48	0.55	0.35	0.69	0.51	0.51
CC_i	**0.590**		**0.615**		**0.602**		**0.600**		**0.560**		**0.634**		**0.580**	
Ranking	**5**		**2**		**3**		**4**		**7**		**1**		**6**	

but also buy (60%). We asked about the B2C websites of the next seven retailers: *eBay*,[5] *Zara*,[6] *Privalia*,[7] *Buy Vip*,[8] *Vente Privee*,[9] *Asos*,[10] and *El Corte Inglés*.[11]

In the light of collected data, we first identified the following latent dimensions and factors (C_i) to consider when assessing e-Service Quality:

- *Utilitarian Quality*:
  - Website Quality: Design (C_1) and Contents (C_2).
  - Offered Service: Guarantee (C_3), Offer (C_4), and Customization (C_5).
  - Security: Payment management (C_6), Privacy (C_7), and Trust (C_8).
- *Hedonic Quality* (C_9).

Then, we applied F-TOPSIS (briefly introduced in Sect. 6.2.3) with the aim of ranking the seven B2C websites under study with respect to the nine C_j factors listed above. Table 6.1 summarizes the results of applying F-TOPSIS on the available data. This table is structured as follows. Columns are related to websites while rows are related to factors. For each website, we report positive (d^+) and negative (d^-) distance between the weighted criteria and ideal solutions. At the bottom, the last two rows show the closeness coefficients CC_i and the final ranking. The B2C website of *Asos* turns up with the highest score (0.634) for e-Service Quality. However, it is closely followed by *Zara* (0.615). Behind them, we find *Privalia*

[5]http://www.ebay.es/.

[6]http://www.zara.com.

[7]http://www.privalia.com/.

[8]http://es.buyvip.com/.

[9]http://www.vente-privee.com/.

[10]http://www.asos.com.

[11]https://www.elcorteingles.es/.

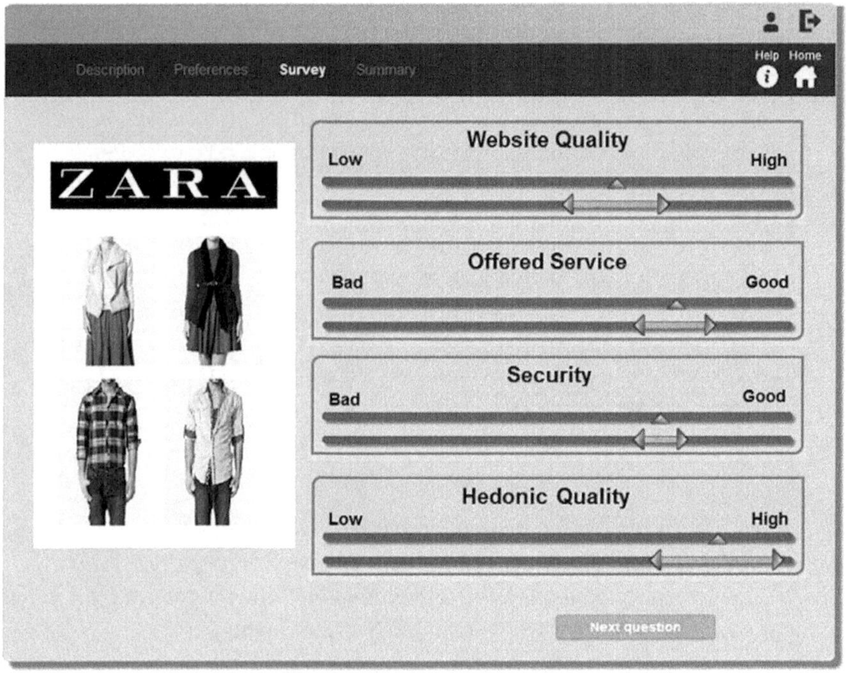

Fig. 6.1 Example of fuzzy rating scale-based questionnaire designed by Quale

(0.602), *Buy Vip* (0.600), *eBay* (0.590), *El Corte Inglés* (0.580), and *Vente Privee* (0.560).

Later, we designed a second survey with the aim of making a finer complementary study regarding the same seven retailers considered previously. This survey was supported by an on-line fuzzy rating scale-based questionnaire (see Fig. 6.1).

We collected data from 78 assessors. They were selected randomly, but respecting the same sample distribution, concerning those assessors who took part in the first study. For each website, assessors had to evaluate four attributes related to the main latent dimensions previously identified: (1) Website Quality, (2) Offered Service, (3) Security, and (4) Hedonic Quality. Notice that the first three attributes are related to the Utilitarian Quality. Each attribute was evaluated in a fuzzy rating scale like the ones depicted in Fig. 6.1. The narrower the triangle support, the more confident the answer is.

Both the design of the fuzzy questionnaire and the analysis of collected data were made as we briefly sketched in Sect. 6.2.2. As result, we obtained a report with, among others, the following contents:

- **Attribute correlation matrix.** We computed Pearson correlation between each pair of attributes under study. Figure 6.2 depicts the correlation matrix in the use case. As expected, the matrix is symmetrical. In addition, correlation is positive

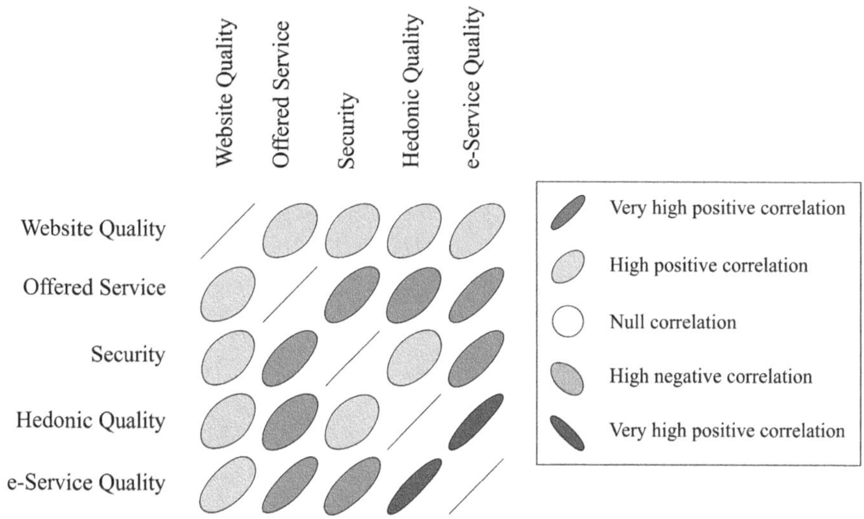

Fig. 6.2 Correlation matrix (Pearson)

in all cases. Moreover, it is easy to appreciate how e-Service Quality is mainly correlated with Hedonic Quality. With respect to the latent factors of Utilitarian Quality, we observe stronger correlation of e-Service Quality with the Offered Service and Security than with the Website Quality.

- **Spider plots.** These graphs summarize at once all collected assessments (in average score) regarding all attributes for a given sample. This fact makes intuitive the comparison among all websites under study (see Fig. 6.3).

  Each attribute is represented by a grey sector. The larger the area of the sector is, the higher the related score. All retailers get high score (above 74) for Website Quality, but the highest score (86) is achieved by *Vente Privee*. Nevertheless, *Vente Privee* gets the lowest score (66.75) regarding Offered Service. In addition, the best service is offered by *Zara* (79.25). From Security point of view, *eBay* is the most appreciated (77) while *Privalia* is the least appreciated (73.5). Notice that security of all websites is considered almost equal.

  As expected, the evaluation of Hedonic Quality exhibits a larger dispersion of answers and a smaller consensus. The highest score (94.5) is achieved by *Asos* while the lowest score (53.75) corresponds to *El Corte Inglés*.

- **Ranking of retailers regarding e-service quality.** Figure 6.4 shows a bar chart with all seven retailers under study. They are ordered in accordance with the average scores computed after processing the data collected in the second survey.

  On the left hand side of the picture, inside the rectangle, we can see bars which correspond to those websites for which assessors were in agreement. Among them, *Zara* (78.25) turns up as the one with the highest e-Service Quality, even though *Asos* (77) is not too far away. The lowest score corresponds to *El Corte Inglés* (although *Privalia* is close). It is worthy to note that we keep on the right

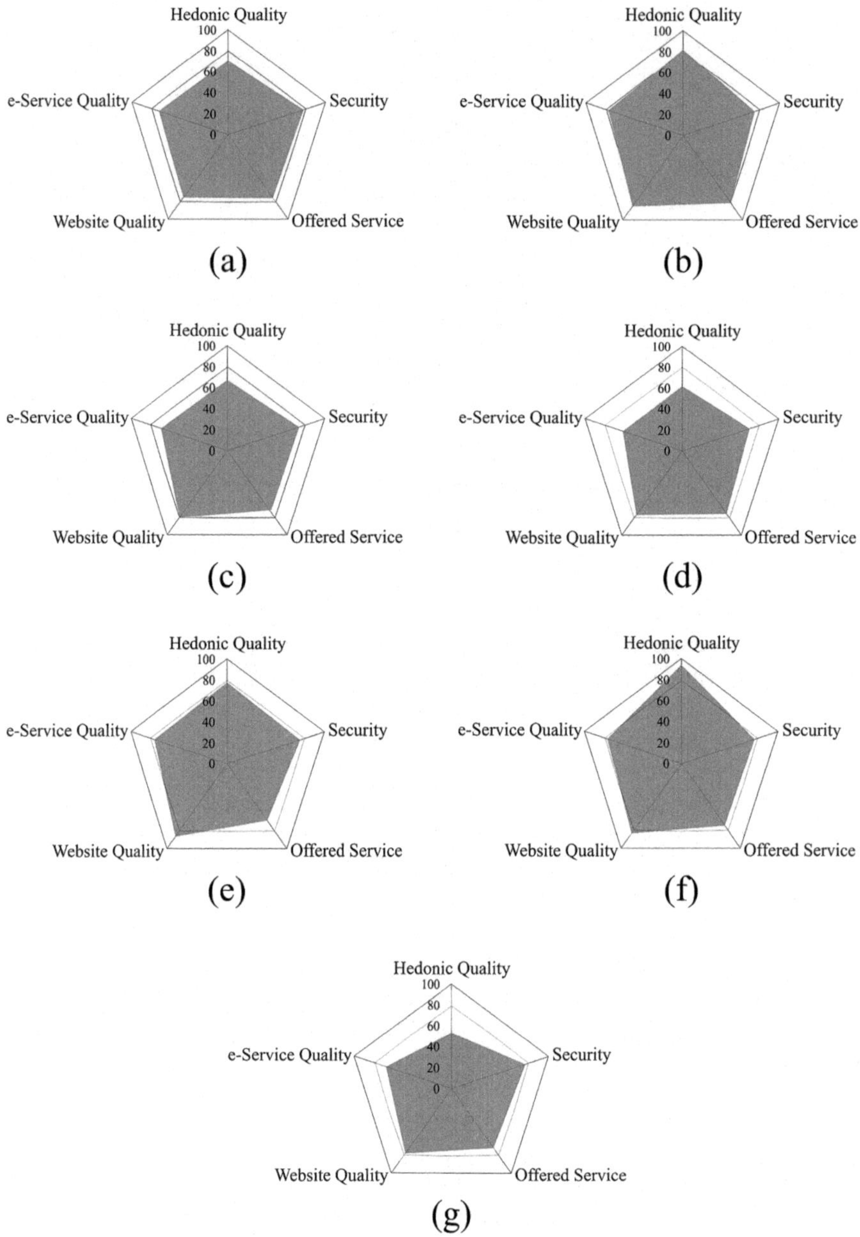

Fig. 6.3 Comparison among B2C websites by spider plots. (**a**) eBay. (**b**) Zara. (**c**) Privalia. (**d**) Buy Vip. (**e**) Vente Privee. (**f**) Asos. (**g**) El Corte Inglés

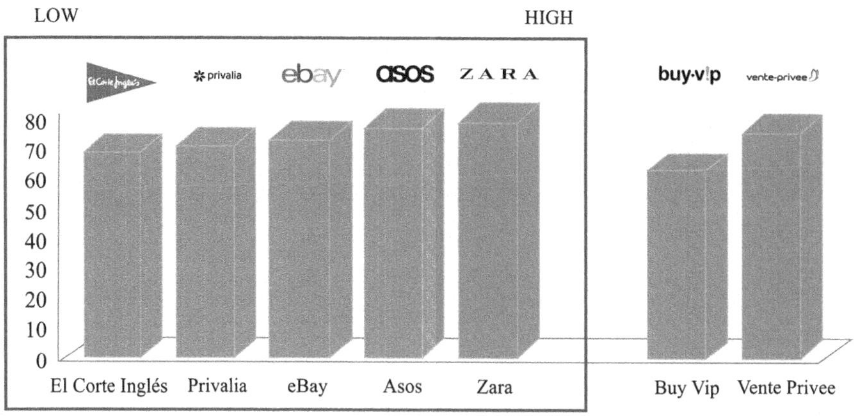

Fig. 6.4 Ranking of retailers provided by Quale with respect to e-service quality

side of the picture, out of the given ranking, the two retailers (*Buy Vip* and *Vente Privee*) for which assessors were not in agreement. So, their related score is not faithful and we must be careful in the comparison against the other retailers.

With the aim of giving a deeper insight with respect to the degree of consensus among assessors, it is needed to take a look at Fig. 6.5. It depicts the distribution of e-Service Quality assessments for all the seven retailers. In each picture, the horizontal axis shows the evaluation range [0,100] while the vertical axis yields the aggregated score normalized in [0,1]. On the one hand, the background curve characterizes all aggregated answers. On the other hand, the foreground bars identify the areas with the greatest answer accumulation. The height of each bar is proportional to the percentage of answers it covers (which is given on top of the bar).

As it can be seen in Fig. 6.5d, there are two disjoint bars to take care in the detailed analysis of e-Service Quality for *Buy Vip*. Moreover, the second bar (16%) is important enough in order not to be ignored. Quale remarks this fact through a warning symbol which is depicted as a triangle with an exclamation mark inside. Anyway, the main bar represents 55% of answers. Therefore, its center of gravity can be seen as a more representative score than the average score for the whole distribution. The situation is even worse in the case of *Vente Privee* (see Fig. 6.5e) where Quale yields a heavy warning (depicted as a double triangle with an exclamation mark inside) because the two bars are really close (39% versus 32%). This means we cannot trust on the aggregated score because we have two plausible values which are likely to yield to two different rankings. Thus, we recommend excluding *Vente Privee* from the final ranking which is as follows: *Zara* (78.25), *Asos* (77), *eBay* (72), *Privalia* (69.5), *El Corte Inglés* (68), and *Buy Vip* (63).

We would like to remark that this ranking is quite similar to the one provided by F-TOPSIS (see Table 6.1) but there are some subtle and valuable differences.

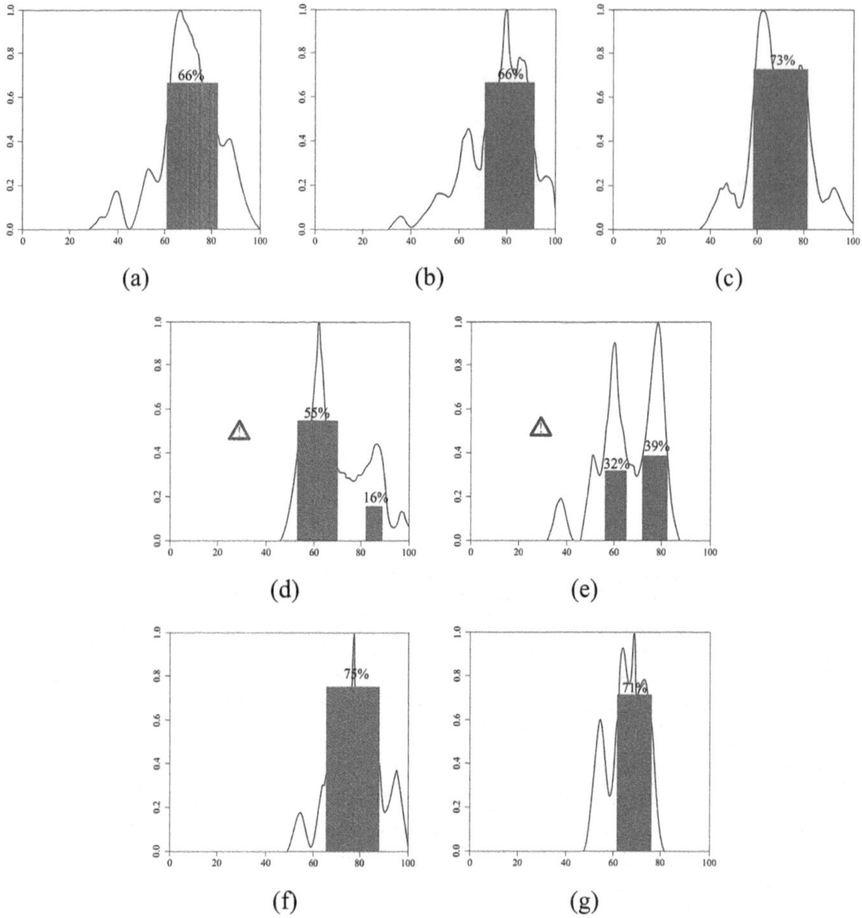

Fig. 6.5 Distribution of collected answers regarding e-service quality. (**a**) eBay. (**b**) Zara. (**c**) Privalia. (**d**) Buy Vip. (**e**) Vente Privee. (**f**) Asos. (**g**) El Corte Inglés

Firstly, both methods place *Zara* and *Asos* at the top of the ranking but with exchanged positions. Anyway, both retailers get so close scores that we can say there is not any difference between them. Secondly, far from the top, *eBay* and *Privalia* turn up also quite close in the middle of both rankings. In addition, *El Corte Inglés* is slightly behind and it goes to the last position in case of excluding the two retailers (*Buy Vip* and *Vente Privee*) which were pointed out by Quale because of the lack of consensus agreement in collected answers. Notice that this important issue is not taken into account by F-TOPSIS. So, we can conclude that Quale helps us to make a finer and deeper analysis than F-TOPSIS.

6.3.2 *e-Service Quality Modeling*

In the light of the analysis made in the previous section, we proposed characterizing e-Service Quality by the model depicted in Fig 6.6. Two latent sub-dimensions of e-Service Quality are observed: (1) Utilitarian Quality and (2) Hedonic Quality.

In addition, there are three latent sub-dimensions (Website Quality; Offered Service; Security) of Utilitarian Quality. They are somehow correlated as it was shown in Fig 6.2. More deeply, it is worthy to note that Website Quality is usually described in terms of website design and contents. In addition, Offered Service depends on guarantee, offer, and customization of service. Security involves payment management, privacy and trust.

We would like to remark once again the fact that evaluations given by users of B2C websites are inherently imprecise and uncertain, as they are based on human perceptions which are inherently subjective. Therefore, the design and implementation of the model introduced above must be made carefully in order to become operative, dynamic and adaptive in nature. Thus, we have implemented the proposed model in the form of a hierarchical fuzzy system with two layers.

Firstly, we addressed the knowledge extraction and representation task from experts. We asked a panel of on-line marketing experts to characterize inputs and outputs as well as relating them through fuzzy IF-THEN rules. Then, these expert KBs were enhanced by adding induced knowledge. We applied data mining tools provided by GUAJE software in order to extract valuable knowledge from data coming out of the second on-line survey on B2C websites which was described in the previous section. The combination of expert and induced knowledge was made by following HILK fuzzy modeling methodology (briefly introduced in Sect. 6.2.4 and implemented by GUAJE). It is worthy to note that the inference process is performed with the usual min-max fuzzy inference mechanism. Moreover, the well-known center of gravity is applied in the defuzzification stage.

We started with setting up a preliminary expert KB to assess e-Service Quality at the top of the hierarchy. It takes two input variables (Utilitarian Quality and Hedonic Quality) and produces one output variable (e-Service Quality). All the three variables are defined by strong fuzzy partitions (see pictures on the left of

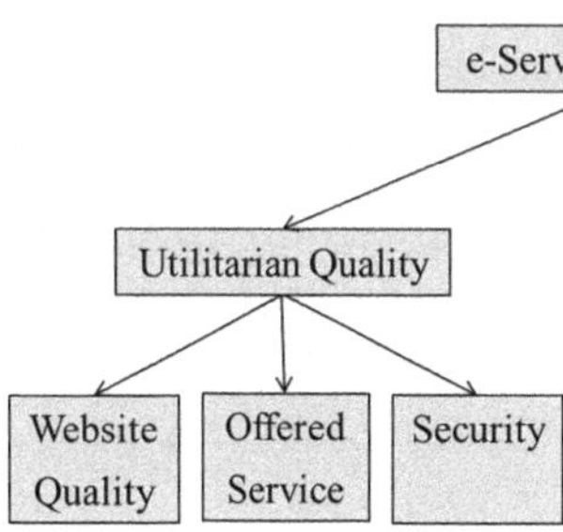

Fig. 6.6 Model for characterizing e-service quality

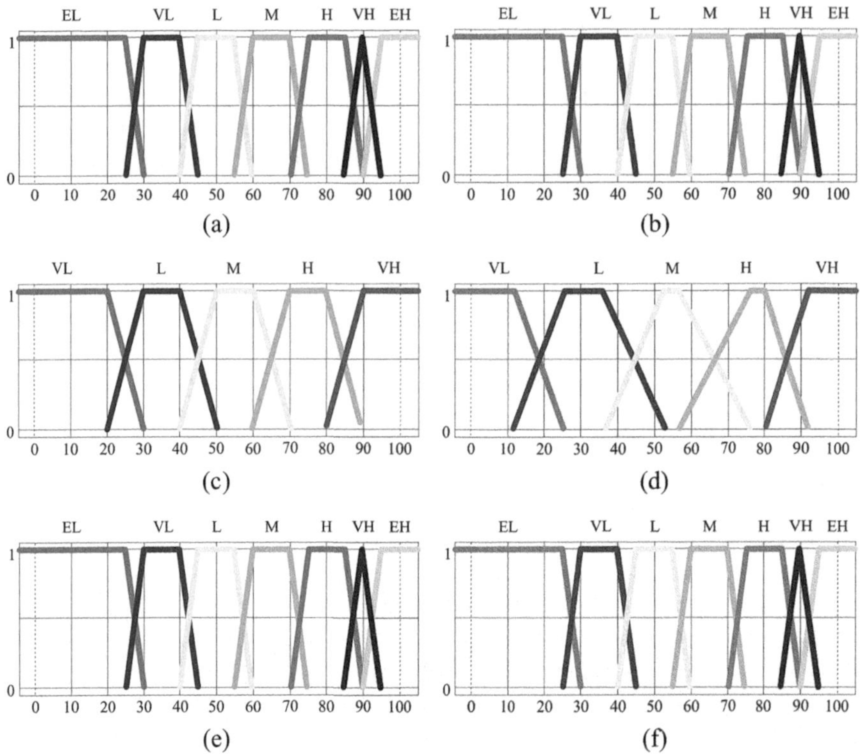

Fig. 6.7 Strong fuzzy partitions. (**a**) Expert utilitarian quality. (**b**) Optimized utilitarian quality. (**c**) Expert hedonic quality. (**d**) Optimized hedonic quality. (**e**) Expert e-service quality. (**f**) Optimized e-service quality

Fig. 6.7). e-Service Quality and Utilitarian Quality are made up of seven fuzzy sets each, with their related linguistic terms: Extremely Low (EL); Very Low (VL); Low (L); Medium (M); High (H); Very High (VH); Extremely High (EH). Hedonic Quality includes only five fuzzy sets (and the set of linguistic terms is a subset of the previous one). As it can be appreciated in Table 6.2, there are 35 expert rules. For example, the first rule can be read as follows "If Utilitarian Quality is Extremely Low and Hedonic Quality is Very Low Then e-Service Quality is Extremely Low".

Then, we applied the data mining tools provided by GUAJE in order to enrich the previous expert KB with knowledge automatically extracted from data. The adjustment of learning parameters and goodness of the designed KB was evaluated through 10-fold cross-validation. For each fold, we first derived rules from a pruned fuzzy decision tree. Secondly, we merged expert and induced rules through a linguistic simplification procedure. Finally, we refined fuzzy partitions by means of the Solis-Wetts tuning mechanism.

On the one hand, regarding training data, coverage measure achieved 100%, mean absolute error (MAE) was 1.47, and root mean square error (RMSE) was

Table 6.2 Expert rules to assess e-service quality in the textile and fashion sector

		Hedonic quality				
		Very Low (VL)	Low (L)	Medium (M)	High (H)	Very High (VH)
Utilitarian quality	Extremely Low (EL)	EL	VL	VL	L	M
	Very Low (VL)	VL	VL	L	M	M
	Low (L)	VL	L	L	M	H
	Medium (M)	VL	L	M	H	H
	High (H)	L	M	M	H	VH
	Very High (VH)	L	M	H	H	EH
	Extremely High (EH)	M	M	H	VH	EH

1.92. On the other hand, coverage arose to 100% while MEA was 1.68 and RMSE was 2.19, with respect to test data. Regarding interpretability indicators, in average, the number of rules was 11.5, the total rule length was 21, and the number of simultaneously fired rules was 2.88 in training and 2.78 in test.

Later, we repeated the same procedure to build the KB in the second layer of the hierarchy. It takes three input variables (Website Quality, Offered Service, and Security) and produces one output variable (Utilitarian Quality). Its goodness was also evaluated through 10-fold cross-validation. To sum up with, coverage was 99.91%, MAE was 2.43, and RMSE was 3.06, with respect to training data; while coverage was 100%, MAE was 2.36, and RMSE was 3.08, with respect to test data. In addition, the number of rules was 7, the total rule length was 12.6, and the number of simultaneously fired rules was 2.9 in training and 2.86 in test.

As a result, the designed model exhibits a good interpretability-accuracy trade-off since it is able to achieve high accuracy with a small set of highly readable linguistic rules. The final model considers all available data in combination with expert knowledge. Pictures on the right hand side of Fig. 6.7 depict the optimized fuzzy partitions. Moreover, the final 11 rules related to e-Service Quality assessment are as follows:

IF *Utilitarian Quality is EL OR VL* AND *Hedonic Quality is L OR M* THEN *e-Service Quality is VL*
IF *Utilitarian Quality is EL OR VL* AND *Hedonic Quality is M OR H* THEN *e-Service Quality is L*
IF *Utilitarian Quality is L* AND *Hedonic Quality is M* THEN *e-Service Quality is L*
IF *Utilitarian Quality is L OR M* AND *Hedonic Quality is H* THEN *e-Service Quality is M*
IF *Utilitarian Quality is M OR H* AND *Hedonic Quality is M* THEN *e-Service Quality is M*
IF *Utilitarian Quality is L OR M OR H* AND *Hedonic Quality is VH* THEN *e-Service Quality is H*
IF *Utilitarian Quality is M OR H OR VH* AND *Hedonic Quality is H* THEN *e-Service Quality is H*
IF *Utilitarian Quality is H OR VH* AND *Hedonic Quality is VH* THEN *e-Service Quality is VH*
IF *Utilitarian Quality is VH OR EH* AND *Hedonic Quality is VH* THEN *e-Service Quality is EH*
IF *Hedonic Quality is VL* THEN *e-Service Quality is VL*
IF *Hedonic Quality is L* THEN *e-Service Quality is L*

Once the proposed fuzzy model was validated, we embedded it in the core of an intelligent virtual assessor able to replicate the evaluations collected through the second survey described in the previous section. In practice, given the numerical values related to all factors defining a website (design, contents, guarantee, and so on), the virtual assessor is able to carry out a fuzzy inference yielding as result a global e-Service Quality score.

This way, the related ranking (with computed scores in brackets) is as follows: (1) Zara [82.5], (2) Asos [75.71], (3) eBay [72.47], (4) Privalia [70.46], (5) El Corte Inglés [69.22], and (6) Buy Vip [62.97]. It is worthy to note that Vente Privee was deliberately excluded from this ranking because, as we explained in the previous section, there was a lack of consensus among collected answers for the related website. Even though there are some minor differences between inferred scores and actual ones, this final ranking is fully in accordance with the one provided in the previous section (see Fig. 6.4). In consequence, the virtual assessor is ready to be used in prospective market research studies with the aim of estimating the e-Service Quality related to other websites different from those considered here; without requiring to disturb consumers with additional surveys.

6.4 Concluding Remarks and Future Work

This paper has presented a novel and efficient methodology for predictive analytics supported by business intelligence tools. We have expanded and further explored the knowledge on e-service quality, addressing a joint application of evaluations on hedonic and utilitarian dimensions by means of combined use of marketing methods (questionnaires) and the Computational Theory of Perceptions (Fuzzy Logic). Moreover, we have applied the paradigm of interpretable fuzzy modeling to deal properly with the uncertainty and imprecision characteristics of human perceptions.

As a result, we have translated sensory data collected through fuzzy rating scale-based questionnaires into valuable knowledge for business decision-making support. Moreover, the interpretability of the designed models is in the core of our human-centric approach. Accordingly, it yields reports easy to understand even by non-experts in the domain of interest as we have proved in a case study regarding B2C websites in the Spanish textile and fashion sector. Reports include several graphs easy to interpret along with a global ranking of retailers regarding e-service quality. Notice that the novel method presented in this paper is able to carry out a finer and deeper analysis than the well-known F-TOPSIS ranking method which we considered for comparison purposes. It is also worthy to remark that the designed virtual assessor is ready to automatically evaluate (without needing to ask directly to consumers) unknown websites out of the seven retailers under study.

In this work we have shown some of the main advantages and drawbacks of our fuzzy approach for e-service quality modeling. Fuzzy sets and systems are well-known because of their ability to properly handle imprecision and uncertainty.

Moreover, we adopted a human-centric modeling approach which yields a good interpretability-accuracy trade-off.

Nevertheless, a lot of work still remains to do. This paper opens the door to very challenging future research. For instance, the use of virtual assessors for reducing costs (mainly time and money) in future market research studies. Also, we plan exploring how to enhance our framework with advanced cloud computing and social network analysis tools.

Finally, let us remark that this work has been developed with the help of several software tools. Please, the interested reader is kindly referred to [1] for further details about them as well as other interesting fuzzy systems software.

Acknowledgements This work is supported by RYC-2016-19802 (Ramón y Cajal contract), and two MINECO projects TIN2017-84796-C2-1-R (BIGBISC) and TIN2014-56633-C3-3-R (ABS4SOW). All of them funded by the Spanish "Ministerio de Economía y Competitividad". Financial support from the Xunta de Galicia (Centro singular de investigación de Galicia accreditation 2016–2019) and the European Union (European Regional Development Fund - ERDF), is gratefully acknowledged.

References

1. Alcala-Fdez, J., & Alonso, J. M. (2016). A survey of fuzzy systems software: Taxonomy, current research trends and prospects. *IEEE Transactions on Fuzzy Systems, 24*(1), 40–56.
2. Alonso, J. M., Castiello, C., & Mencar, C. (2015). Interpretability of fuzzy systems: Current research trends and prospects. In *Handbook of computational intelligence* (pp. 219–237). Berlin: Springer.
3. Alonso, J. M, Castiello, C., & Mencar, C. (2018). A bibliometric analysis of the explainable artificial intelligence research field. In *Information Processing and Management of Uncertainty in Knowledge-Based Systems - Theory and Foundations*. Communications in computer and information science (Vol. 853, pp. 3–15). Cham: Springer. https://doi.org/10.1007/978-3-319-91473-2_1
4. Alonso, J. M., & Magdalena, L. (2011). HILK++: An interpretability-guided fuzzy modeling methodology for learning readable and comprehensible fuzzy rule-based classifiers. *Soft Computing, 15*, 1959–1980.
5. Alonso, J. M., Pancho, D. P., Magdalena, L., Nunez, D. A., Sanchez, D. S., Suarez, P. F., et al. (2015). QUALE®: A new toolbox for quantitative and qualitative analysis of human perceptions. In *Proceedings of 16th World Congress of the International Fuzzy Systems Association (IFSA) 9th Conference of the European Society for Fuzzy Logic and Technology (EUSFLAT)* (pp. 659–666). Paris: Atlantis Press.
6. Carifio, J., & Perla, R. (2007). Ten common misunderstandings, misconceptions, persistent myths and urban legends about Likert scales and Likert response formats and their antidotes. *Journal of Social Sciences, 3*(3), 106–116.
7. Casillas, J., Cordon, O., Herrera, F., & Magdalena, L. (2003). *Interpretability issues in fuzzy modeling. Studies in fuzziness and soft computing*. Berlin: Springer.
8. Castro-Lopez, A., Puente, J., & Vazquez-Casielles, R. (2017). Fuzzy inference suitability to determine the utilitarian quality of B2C websites. *Applied Soft Computing, 57*, 132–143.
9. Chen, C. T., Lin, C. T., & Huang, S. F. (2006). A fuzzy approach for supplier evaluation and selection in supply chain management. *International Journal of Production Economics, 102*, 289–301.

10. CNMC. (2017). *Second quarter 2017 report on Spanish e-commerce*. Accessed May 2018.
11. de la Rosa de Saa, S., Gil, M. A., Gonzalez-Rodriguez, G., Lopez, M. T., & Lubiano, M. A. (2015). Fuzzy rating scale-based questionnaires and their statistical analysis. *IEEE Transactions on Fuzzy Systems, 23*, 111–126.
12. Gil, M. A., Lubiano, M. A., de la Rosa de Saa, S., & Sinova, B. (2015). Analyzing data from a fuzzy rating scale-based questionnaire. A case study. *Psicothema, 27*(2), 182–191.
13. Kabir, G., Sadiq, R., & Tesfamariam, S. (2014). A review of multi-criteria decision-making methods for infrastructure management. *Structure and Infrastructure Engineering, 10*, 1176–1210.
14. Likert, R. (1932). A technique for the measurement of attitudes. *Archives of Psychology, 140*, 1–55.
15. Pancho, D. P., Alonso, J. M., & Magdalena, L. (2013). Quest for interpretability-accuracy trade-off supported by fingrams into the fuzzy modeling tool GUAJE. *International Journal of Computational Intelligence Systems, 6*(Suppl. 1), 46–60.
16. Parasuraman, A., Zeithaml, V., & Malhotra, A. (2005). E-S-Qual: A multiple-item scale for assessing electronic service quality. *Journal of Service Research, 7*(10), 1–21.
17. Posner, H. (2016). *Fashion marketing*. Barcelona: Gustavo Gili Editorial.
18. Quirós, P., Alonso, J. M., & Pancho, D. P. (2016). Descriptive and comparative analysis of human perceptions expressed through fuzzy rating scale-based questionnaires. *International Journal of Computational Intelligence Systems, 9*(3), 450–467.
19. Rares, O. (2014). Measuring perceived service quality offline vs. online: A new PeSQ conceptual model. *Procedia Economics and Finance, 15*, 538–551.
20. Salehi, M., Salimi, M., & Haque, A. (2013). The impact of online customer experience (OCE) on service quality in Malaysia. *Word Applied Sciences Journal, 21*(11), 1621–1631.
21. Sun, C. C., & Lin, G. T. R. (2009). Using fuzzy TOPSIS method for evaluating the competitive advantages of shopping websites. *Expert Systems with Applications, 36*(9), 11764–11771.
22. The European Apparel and Textile Confederation. (2016). *Taking action for the future of the European textile and clothing industry*. Accessed May 2018.
23. Wang, Y. J., & Lee, H. (2007). Generalizing TOPSIS for fuzzy multiple-criteria group decision-making. *Computers and Mathematics with Applications, 53*, 1762–1772.
24. Zadeh, L. A. (2001). A new direction in AI: Toward a computational theory of perceptions. *Artificial Intelligence Magazine, 22*(1), 73–84.
25. Zeng, X., Ruan, D., & Koehl, L. (2008). Intelligent sensory evaluation: Concepts, implementations and applications. *Mathematics and Computers in Simulation, 77*, 443–452.

Chapter 7
Grey Number Based Methodology for Non-homogeneous Preference Elicitation in Fuzzy Risk Analysis Management

Ahmad Syafadhli Abu Bakar, Ku Muhammad Naim Ku Khalif, Abdul Malek Yaakob, Alexander Gegov, and Ahmad Zaki Mohamad Amin

7.1 Introduction

The incomplete and vagueness of real-world information has triggered the emergence of grey system in human decision making environment. The grey system serves as an alternative methodology that plays the role in complementing the uncertainty in systems with partial information [1–4]. Similarly as fuzzy sets [5] and rough sets [6–8], grey sets characterised the uncertainties in the form of grey

A. S. A. Bakar (✉)
Mathematics Division, Centre for Foundation Studies in Science, University of Malaya, Kuala Lumpur, Malaysia

Centre of Research for Computational Sciences and Informatics in Biology, Bioindustry, Environment, Agriculture and Healthcare (CRYSTAL), University of Malaya, Kuala Lumpur, Malaysia
e-mail: ahmadsyafadhli@um.edu.my

K. M. N. K. Khalif
Department of Science Program (Mathematics), Faculty of Industrial Sciences and Technology, Universiti Malaysia Pahang, Pahang, Malaysia
e-mail: kunaim@ump.edu.my

A. M. Yaakob
School of Quantitative Sciences, Universiti Utara Malaysia, Kedah, Malaysia
e-mail: abd.malek@uum.edu.my

A. Gegov
School of Computing, University of Portsmouth, Portsmouth, UK
e-mail: alexander.gegov@port.ac.uk

A. Z. M. Amin
Mathematics Division, Centre for Foundation Studies in Science, University of Malaya, Kuala Lumpur, Malaysia
e-mail: azaki@um.edu.my

A. Meier et al. (eds.), *Applying Fuzzy Logic for the Digital Economy and Society*, Fuzzy Management Methods, https://doi.org/10.1007/978-3-030-03368-2_7

numbers as the basic concept in grey systems [9, 10]. A grey number is defined as a number with an unknown position within clear lower and upper boundaries [3, 9]. The main aim of introducing grey numbers in the literature is to define the membership or characteristic function value that is unclear in traditional crisp sets and fuzzy sets [3, 9].

Membership or characteristic function values are often used in decision making process as the preferences elicited by decision makers. However, determination of a suitable preference elicitation for a situation is not an easy task as different decision makers may have different types of perception. For instance, in real risk analysis world scenarios, it is a big challenge for risk analysts to make a proper and comprehensive decision when coping with the risks. This is because different risk analysts may mitigate the level of harm of the same risk differently. Another major concern in many practical risk analysis problems is they do not have flexibility with regards to knowledge elicitation and disagreements in the group. This is due to the non-homogeneous nature of risk analysts' preferences that lead to inconsistent agreements in the process of group decision making. Thus, the element of non-homogeneous in the membership or characteristic function values is important to be addressed in order to complement the non-homogeneous nature of risk analysts' preferences.

In the literature, there are many established concepts that are also concerned with the study on membership or characteristic function value such as rough sets, type-2 fuzzy sets and interval-valued fuzzy sets [11, 12]. Nonetheless, all of them have weaknesses from one to another. Rough sets have successfully expressed this situation by representing the probability of an element being a member of the set using rough membership function [6–8]. However, the representation is incomplete when some well-defined values that belong to the decision making situations are missing. Type-2 fuzzy sets [13] on the other hand, define the membership value using another fuzzy set which includes the Footprint of Uncertainty [14, 15]. Nevertheless, it is difficult to clarify one fuzzy set with another fuzzy set [10] due to the fact that the uncertain membership value needs a representation that can express both possible values of type-2 fuzzy sets.

More importantly, the value is a single value as defined in fuzzy sets. Interval-valued fuzzy sets conceptually solve this issue in the case of fuzzy sets when grey sets are considered to be the same as interval-valued fuzzy sets. This is due to grey numbers and intervals shared some common aspects [16]. Nonetheless, this understanding is a misconception, as grey numbers have special features in which intervals do not have. In addition, this concept is inconsistent with respect to the epistemic uncertainty of an interval representation. Furthermore, grey sets provide better coverage when dealing with partial information than interval-valued fuzzy sets [10].

As grey numbers [3, 9] are capable to efficiently describe non-homogeneous membership or characteristic function values [10], numerous efforts in the literature have adopted and applied grey numbers towards decision making problems. Among others are [17] in supply chain management model, forecasting [18], software effort estimation model [19], grey-TOPSIS in subcontractor selection [20] and contractor's selection [21]. Nevertheless, these applications have shortcomings and drawbacks because they conceptually utilised the aforementioned established concepts that are proven to be inconsistent with grey numbers.

As human preferences elicitation are non-homogeneous in nature, utilisation of grey numbers provides better representations for human related decision making. Thus, to complement both theoretical methodology and decision making application of grey numbers, this paper proposes a novel non-homogeneous preference elicitation based on grey numbers for risk analysis problem. This work also introduces a novel theoretical non-homogeneous consensus reaching method that resolves disagreement between risk analysts. A novel decision making approach that is developed based on the ranking concept, is then introduced to complement the consensus reaching method in solving decision making problems involving grey numbers. Later on, validations on both novelties are presented along with real world case study, as to demonstrate the novelty, validity and feasibility of the proposed methodology.

The rest of the paper is structured as follows. Section 7.2 provides brief overviews on theoretical preliminaries related to this study. Section 7.3 discusses the relevance of grey numbers in risk analysis management. Section 7.4 presents the research methodology of this study. Section 7.5 covers validation of results obtained throughout this study. Section 7.6 concerns with the application of the research methodology on real world case study and finally, the conclusion is given in Sect. 7.7.

7.2 Theoretical Preliminaries

7.2.1 Fuzzy Number

Definition 7.2.1 ([22]) A triangular type-1 fuzzy number A is represented by Eq. (7.1).

$$\mu_A(x) = (a_1, a_2, a_3; 1) = \begin{cases} \frac{(x-a_1)}{(a_2-a_1)} & \text{if } a_3 \leq x \leq a_4 \\ \frac{(x-a_4)}{(a_3-a_4)} & \text{if } a_3 \leq x \leq a_4 \\ 0 & \text{otherwise} \end{cases} \tag{7.1}$$

Definition 7.2.2 ([22]) A trapezoidal type-1 fuzzy number A is represented by Eq. (7.2).

$$\mu_A(x) = (a_1, a_2, a_3, a_4; 1) = \begin{cases} \frac{(x-a_4)}{(a_3-a_4)} & \text{if } a_2 \leq x \leq a_3 \\ 1 & \text{if } a_2 \leq x \leq a_3 \\ \frac{(x-a_4)}{(a_3-a_4)} & \text{if } a_3 \leq x \leq a_4 \\ 0 & \text{otherwise} \end{cases} \tag{7.2}$$

7.2.2 Grey Number

Definition 7.2.3 ([10]) A grey number, G_A, is a number with clear upper and lower boundaries but has an unknown position within the boundaries. Mathematically, a grey number for the system is expressed as:

$$G_A \in \lfloor g^-, g^+ \rfloor = \{g^- \leq t \leq g^+\} \tag{7.3}$$

where t is information about $g^{\pm}$ while g^- and g^+ are the upper and lower limits of information t respectively.

As mentioned in the introduction section, grey number is introduced in the literature as to clearly define the membership or characteristic function values of a set. Therefore, in this paper, the terms grey number is used interchangeably with characteristic function value and vice versa.

Definition 7.2.4 ([10]) For a set $A \subseteq U$, if its characteristic function value of each x with respect to A, $g_A^{\pm}(x)$, can be expressed with a grey number, $g_A^{\pm}(x) \in \bigcup_{i=1}^{n}[a_i^-, a_i^+] \in D[0,1]^{\pm}$, then A is a grey set, where $D[0,1]^{\pm}$ is the set of all grey numbers within the interval [0, 1].

In the literature on grey numbers, if the value of the characteristic function is completely known or completely unknown, then it is called as the white number or black number respectively. In other words, characteristic function value 1 refers to the element is a white numbers and 0 is a black number. Likewise, any values in [0, 1] are considered as the grey numbers. Consider the following definitions by [10].

Definition 7.2.5 (White Sets) For a set $A \subseteq U$, if its characteristic function value of each x with respect to A, $g_{A_i}^{\pm}$, $i = 1, 2, \ldots, n$, can be expressed with a white number, then A is a white set.

Definition 7.2.6 (Black Sets) For a set $A \subseteq U$, if its characteristic function value of each x with respect to A, $g_{A_i}^{\pm}$, $i = 1, 2, \ldots, n$, can be expressed with a white number, then A is a black set.

Definition 7.2.7 (Grey Sets) For a set $A \subseteq U$, if its characteristic function value of each x with respect to A, $g_{A_i}^{\pm}$, $i = 1, 2, \ldots, n$, can be expressed with a white number, then A is a grey set.

Definition 7.2.8 Let U be the finite universe of discourse, x be an element and $x \in U$. For a grey set $A \subseteq U$, the characteristic function value of x with respect to A is $g_A^{\pm}(x) \in D[0,1]^{\pm}$. The degree of greyness, $g_A^o(x)$, of element x for set A is expressed as

$$G_A \in \lfloor g^-, g^+ \rfloor = \{g^- \leq t \leq g^+\} \tag{7.4}$$

Definition 7.2.9 (Degree of Greyness of a Set[10]) Let U be the finite universe of discourse, A be a grey set and $A \subseteq U$. x_i is element relevant to A and $x_i \in U$ $i = 1, 2, \ldots, n$ and n is the cardinality of U. The degree of greyness of set A, g_A^*, is defined as

$$g_A^* = \frac{\sum_{i=1}^{n} g_A^o(x_i)}{n} \tag{7.5}$$

It is worth pointing out here that Eq. (7.5) can be expressed in term of fuzzy set expression [10], given by

$$A = g_A^{\pm}/x_1 + g_A^{\pm}/x_2 + \ldots + g_A^{\pm}/x_n \tag{7.6}$$

7.3 Relevance of Grey Numbers in Risk Analysis Management

In this section, a case study on risk analysis problem is carried out as to demonstrate the relevance of grey numbers towards non-homogeneous preference elicited by risk analyst. Information on the case study is summarised in Table 7.1, given as follows.

In Table 7.1, criteria B and C for each company under consideration are preferences elicited by risk analyst 1. It is also noted that risk level, D, which is defined based on criteria B and C is also in the form of preference elicitation. These preferences elicitation are expressed into characteristic functions defined as Eqs. (7.7) and (7.8) for B & D and C & D respectively.

$$f_{CD}(A_i) = \begin{cases} 1 & \text{if C= high} \\ [0,1] & \text{if C= medium} \\ 0 & \text{if C= low} \end{cases} \tag{7.7}$$

$$f_{BD}(A_i) = \begin{cases} 1 & \text{if B= high} \\ [0,1] & \text{if B= medium} \\ 0 & \text{if B= low} \end{cases} \tag{7.8}$$

Table 7.1 Information on risk level evaluation for companies in Malaysia by risk analyst 1

	Criteria		
Company, A	Probability of failure, B	Severity of loss, C	Risk level, D
A_1	B_{A_1} = Low	C_{A_1} = Low	D_{A_1} = Low
A_2	B_{A_2} = Medium	C_{A_2} = Low	D_{A_2} = Medium
A_3	B_{A_3} = Low	C_{A_3} = Medium	D_{A_3} = Low
A_4	B_{A_4} = High	C_{A_4} = High	D_{A_4} = High

From Eqs. (7.7) and (7.8), the following aggregated expressions are obtained.

$$\begin{aligned} B^* &= [0,0]/A_1 + [0,1]/A_2 + [0,0]/A_3 + [1,1]/A_4 \\ &= 0/A_1 + [0,1]/A_2 + 0/A_3 + 1/A_4 \end{aligned} \tag{7.9}$$

$$\begin{aligned} C^* &= [0,0]/A_1 + [0,0]/A_2 + [0,1]/A_3 + [1,1]/A_4 \\ &= 0/A_1 + 0/A_2 + [0,1]/A_3 + 1/A_4 \end{aligned} \tag{7.10}$$

where B^* and C^* are aggregated relationships for B & D and C & D respectively.

It is worth noting here that all preferences elicitation are now in the form of characteristic function values with 0 is the black number, 1 is the white number and is the grey numbers. In other words, the non-homogeneous preferences elicitation expressed by risk analyst 1 are in the form of grey numbers. Although, it is acknowledged based on Definition 7.2.4 that black numbers, white numbers and grey numbers are considered as grey numbers, all of them are still distinct in term of their value forms. Thus, this study describes grey numbers into two value form namely the numerical value and interval value forms. The following Table 7.2 presents details of these value forms of grey numbers.

Descriptions presented in Table 7.2 are important to be introduced here because they point out the non-homogeneous nature of a grey number. Unlike the established research concepts mentioned in the introduction, only one value form (homogeneous) is considered in their computation works that is either numerical value form or interval value form. The non-homogeneous value forms of grey numbers described here indicate that grey numbers are more relevant than established research concepts because both elements and the sets can simultaneously be non-homogeneous in certain decision making problems, for instance Eqs. (7.9) and (7.10). Even though, the significant nature of non-homogeneous value forms of grey numbers creates another level of complexity in terms of the computational methodology works, this challenge brings the motivation of this study.

Table 7.2 Descriptions of grey numbers value forms

Grey number	Value form	Example	
		Equation (7.9)	Equation (7.10)
0	Numerical	$0/A_1, 0/A_3$	$0/A_1, 0/A_2$
[0, 1]	Interval	$[0, 1]/A_2$	$[0, 1]/A_3$
1	Numerical	$1/A_4$	$1/A_4$

7.4 Research Methodology

In this section, novel theoretical methodology to deal with grey numbers is presented. It is worth mentioning here that this methodology consists of two layers namely the consensus reaching method as Layer 1 and the ranking approach as Layer 2. Details on both layers are explained as follows.

7.4.1 Layer 1: Consensus Reaching Method

As mentioned in Sect. 7.3, the value forms of grey number are non-homogeneous (i.e. numerical value form and interval value form). Due to this reason, a novel consensus reaching method which is the conversion of grey numbers into type-1 fuzzy numbers is proposed. The main purpose of the consensus reaching method is to ensure that both value forms of grey numbers are transformed into common value form for easier computation. Furthermore, type-1 fuzzy numbers are well established in decision making application [23–32]. This consensus reaching method is basically an extension of [10] research work on replacing the characteristic function on grey set with fuzzy membership function. Discussions on the aforementioned replacement are given in Sect. 7.5 while details on the consensus reaching method are as follows.

Numerical Value Form
If $g_A^{\pm} \in [0, 1]$ is a numerical value, then $g_A^{\pm}$ is converted into grey type-1 fuzzy numbers using conversion function, T_{1i}, given as follows.

Definition 7.4.1 A numerical value of $g_A^{\pm}$ is converted into grey triangular type-1 fuzzy numbers using conversion function, T_{1i} as

$$\begin{aligned} &T_{1i} : g_A^{\pm} \rightarrow G_A(x) \\ &T_{11} = G_A(x) = (g_{a_1}, g_{a_2}, g_{a_3}) \end{aligned} \tag{7.11}$$

and grey trapezoidal type-1 fuzzy number using conversion function, T_{12} as

$$T_{12} = G_A(x) = (g_{a_1}, g_{a_2}, g_{a_3}, g_{a_4}) \tag{7.12}$$

Interval Value Form
If $g_A^{\pm} \in [0, 1]$ is an interval value, then $g_A^{\pm}$ is converted into grey type-1 fuzzy numbers using conversion function, T_{2i}, given as follows.

Definition 7.4.2 An interval value of $g_A^{\pm}$ is converted into triangular grey type-1 fuzzy numbers using conversion function, T_{2i}:

$$\begin{aligned} &T_{21} : [a, b] \rightarrow G_A(x) \\ &T_{21}[a, b] = G_A(x) = (g_{a_1}, g_{a_2}, g_{a_3}) \end{aligned} \tag{7.13}$$

and grey trapezoidal type-1 fuzzy number using conversion function, T_{22} as

$$T_{22}[a, b] = G_A(x) = (g_{a_1}, g_{a_2}, g_{a_3}, g_{a_4}) \quad (7.14)$$

7.4.2 Layer 2: Ranking Approach

In this subsection, a ranking approach for grey type-1 fuzzy numbers is presented. The complete procedure is given as follows.

Let $G_A(x) = (g_{a_1}, g_{a_2}, g_{a_3}, g_{a_4})$ be a grey type-1 fuzzy number obtained from the conversion approach presented in Sect. 7.4.1. The complete theoretical procedure for ranking grey type-1 fuzzy number is as follows.

Step 1: Calculate the centroid $-x$ value for $G_A(x)$ based on [33] as

$$x_{G_A} = \frac{\int_{-\infty}^{\infty} x f(x) dx}{\int_{-\infty}^{\infty} f(x) dx}$$

and the centroid $-y$ value for $G_A(x)$ as

$$y_{G_A} = \frac{\int_0^{w_{G_A}} \alpha \mid G_A^{\alpha} \mid d\alpha}{\int_0^{w_{G_A}} \mid G_A^{\alpha} \mid d\alpha}$$

where $x_{g_A} \in [0, 1]$ and $y_{g_A} \in [0, 1]$.

Step 2: Compute the spread value for $G_A(x)$ based on [25] as

$$s_{G_A} = i_{G_A} \times ii_{G_A}$$

where $i_{G_A} = \mid g_{a_4} - g_{a_1}$ and $ii_{G_A} = y_{G_A}$

Step 3: Evaluate the ranking value for all grey type-1 fuzzy numbers under consideration as

$$\phi_{G_A} = x_{G_A} \times y_{G_A} \times (1 - s_{G_A}) \quad (7.15)$$

Ranking descriptions:

$$\text{If } \phi_{G_A} > \phi_{G_B}, \text{ then } G_A(x) \succ G_B(x)$$
$$\text{If } \phi_{G_A} = \phi_{G_B}, \text{ then } G_A(x) \approx G_B(x)$$
$$\text{If } \phi_{G_A} < \phi_{G_B}, \text{ then } G_A(x) \prec G_B(x)$$

It is worth mentioning here that the ranking approach presented in this subsection is similar as in [25]. The distinction between [25] and this proposed work is the former is developed for type-1 fuzzy numbers while the latter is purposely made for grey type-1 fuzzy numbers.

7.5 Validation of Results

This section covers validation on the proposed methodology in the previous section. It is worth mentioning here that the relevant properties considered in this section justify the consistency of the proposed extension within the domain of grey numbers and these properties can be extended further.

7.5.1 Layer 1: Consensus Reaching Method

As mentioned in Sect. 7.4.1, the consensus reaching method developed is an extension of [10] work. The following Theorem 7.5.1 justifies the consistency on replacing the characteristic function of grey numbers with fuzzy membership function.

Theorem 7.5.1 *Let U be the finite universe of discourse, A be a grey set and $A \subseteq U$. x is an element and $x \in U$, $g_A^{\pm}(x)$ is the characteristic function value of x with respect to A, $g_A^{o}(x)$ is the degree of greyness of $g_A^{\pm}(x)$ and g_A^{*} is the degree of greyness for A.*

Proposition 3 *A is a type-1 fuzzy set if and only if $g_A^{*} = 0$ and $g_A^{\pm}(x)$ for any $x \in U$*

Proof 2 If A a type-1 fuzzy set, then $g_A^{*} = 0$ and $g_A^{\pm}(x) \in [0, 1]$ for any $x \in U$

Let A be a type-1 fuzzy set expressed as

$$A = \mu_A(x_1)/x_1 + \mu_A(x_2)/x_2 + \ldots + \mu_A(x_n)/x_n \tag{7.16}$$

where $\mu_A(x)$ is the membership degree for A with $\mu_A(x) \in [0, 1]$.

When $\mu_A(x) = g_A^{\pm}(x) \in [0, 1]$, then the following is obtained based on Eq. (7.5).

$$g_A^{*} = \frac{\mid (\mu_A(x_1) - \mu_A(x_1)) + (\mu_A(x_2) - \mu_A(x_2)) + \ldots + (\mu_A(x_n) - \mu_A(x_n)) \mid}{n} = 0 \tag{7.17}$$

where $\mu_A(x) = g_A^{\pm}(x) \in [0, 1]$ for any $x \in U$.

Proposition 4 *If $g_A^{*} = 0$ and $g_A^{\pm}(x) \in [0, 1]$ for any $x \in U$, then A is a type-1 fuzzy set.*

Let A be grey set expressed as

$$A = g_A^{\pm}(x_1)/x_1 + g_A^{\pm}(x_2)/x_2 + \ldots + g_A^{\pm}(x_n)/x_n$$

Based on Definition 7.2.8, $g_A^{\pm}(x_i) \in [0, 1]$ where $i = 1, 2, \ldots, n$, is a single grey number. Thus, the following is hold.

$$g_A^* = \frac{|\,(g_A^{\pm}(x_1) - g_A^{\pm}(x_1)) + (g_A^{\pm}(x_2) - g_A^{\pm}(x_2)) + \ldots + (g_A^{\pm}(x_i) - g_A^{\pm}(x_i))\,|}{n} = 0 \tag{7.18}$$

If $\mu(x) = g_A^{\pm}(x) \in [0, 1]$, then Eq. (7.6) is defined as

$$A = \mu_A(x_1)/x_1 + \mu_A(x_2)/x_2 + \ldots + \mu_A(x_n)/x_n$$

Theorem 7.5.1 holds.

With respect to the novel conversion methodology developed in Sect. 7.4, detail validation is as follows.

Let G_A and μ_A be the grey number and membership value for A respectively, where $G_A \in D[0, 1]^{\pm}$ and $\mu_A \in [0, 1]$.

Numerical Value

Property 1 If $G_A = \mu_A$, then $\mu_A : U \to D[0, 1]^{\pm}$.

Proof 3

$$\begin{aligned} &G_A = \mu_A, \textit{ implies that } G_A = \mu_A \in [0, 1]^{\pm} \\ &\text{hence, } \mu_A : U \to D[0, 1]^{\pm} \text{ (proven)} \end{aligned} \tag{7.19}$$

It is worth noting here that, Eq. (7.19) is consistent with Eqs. (7.16)–(7.18).

Interval Value

Property 2 If membership interval, $t = \lfloor g^-, g^+ \rfloor$, then $\mu : U \to D[0, 1]^{\pm}$.

Proof 4 $t = \lfloor g_A^-, g_A^+ \rfloor$ implies that $t \in D[0, 1]^{\pm}$

For continuous grey numbers, $G_A \in t$, any unknown value of G_A within t indicates that $G_A \in D[0, 1]^{\pm}$. Thus, when $G_A = \mu_A$ then $\mu_A : D \to D[0, 1]^{\pm}$ (proven).

7.5.2 Layer 2: Ranking Approach

Let G_A and G_B be any grey type-1 fuzzy numbers. All ranking properties presented here are based on [34, 35] on ranking fuzzy quantities.

Ranking Property 7.5.1 *If $G_A \succeq G_B$ and $G_B \succeq G_A$, then $G_A \approx G_B$.*

Proof 5 $G_A \succeq G_B$ implies that $\phi_{G_A} \geq \phi_{G_B}$ and $G_B \succeq G_A$ implies that $\phi_{G_B} \geq \phi_{G_A}$, thus $\phi_{G_A} = \phi_{G_B}$ which is $G_A \approx G_B$.

Ranking Property 7.5.2 *If $G_A \succeq G_B$ and $G_B \succeq G_C$, then $G_A \succeq G_C$.*

Proof 6 $G_A \succeq G_B$ implies that $\phi_{G_A} \geq \phi_{G_B}$ and $G_B \succeq G_C$ implies that $\phi_{G_B} \geq \phi_{G_C}$, thus $\phi_{G_A} = \phi_{G_C}$ which is $G_A \succeq G_C$.

Ranking Property 7.5.3 *If $G_A \bigcap G_B = \phi$ and G_A is on the right side of G_B, then $G_A \succeq G_B$*

Proof 7 $G_A \bigcap G_B = \phi$ and G_A is on the right side of G_B implies that $\phi_{G_A} \geq \phi_{G_B}$, thus $G_A \succeq G_B$

Ranking Property 7.5.4 *The order of G_A and G_B are not affected by other grey type-1 fuzzy numbers under comparison.*

Proof 8 The ordering of G_A and G_B are completely determined by ϕ_{G_A} and ϕ_{G_B} respectively, thus the ordering of G_A and G_B are not affected by other grey type-1 fuzzy numbers under comparison.

7.6 Case Study

In this section, assessments on the level of risk of three distinct companies in Malaysia are conducted. It is worth mentioning here that all companies under consideration are of same nature as they are producing the same product. Details on descriptions of severity of loss and probability of failure for each company under consideration in the form of grey numbers are summarised in Table 7.3.

As the methodology developed in Sect. 7.4 consists of two layers, the assessment of level of risk for each company under consideration follows the two layers developed.

Table 7.3 Descriptions of risk assessment of companies in the form of grey numbers

Company	Component	Severity of loss	Probability of failure
C_1	A_{11}	W_{11} = low	S_{11} = fairly-low
	A_{12}	W_{12} = fairly-high	S_{12} = medium
	A_{13}	W_{13} = very-low	S_{13} = fairly-high
C_2	A_{21}	W_{21} = low	S_{21} = very-high
	A_{22}	W_{22} = fairly-high	S_{22} = fairly-high
	A_{23}	W_{23} = very-low	S_{23} = medium
C_3	A_{31}	W_{31} = low	S_{31} = fairly-low
	A_{32}	W_{32} = fairly-high	S_{32} = high
	A_{33}	W_{33} = very-low	S_{33} = fairly-high

7.6.1 Layer 1: Consensus Reaching Method

Based on Table 7.2, it is acknowledged that grey numbers can exist in numerical and interval value forms. Thus, all grey numbers in Table 7.3 are converted into trapezoidal type-1 fuzzy numbers using Eqs. (7.12) and (7.14), as to ensure they are consistent in nature. The complete descriptions on the converted grey numbers into trapezoidal type-1 fuzzy numbers are presented in Table 7.4.

As to complete this layer, details in Table 7.4 are aggregated for consensus reaching purposes. Table 7.5 presents consensus reached for all companies under consideration after aggregating process.

7.6.2 Layer 2: Ranking Approach

Based on Sect. 7.4, details on centroid point, spread and ranking value for each company considered are evaluated and summarised in Table 7.6.

From Table 7.6, it can be concluded that the most risky company is C_2, followed by C_1 and C_3.

Table 7.4 Descriptions of risk assessment of companies in the form of type-1 fuzzy numbers

Company	Component	Severity of loss	Probability of failure
C_1	A_{11}	$W_{11} = (0.04, 0.10, 0.18, 0.23; 1.0)$	$S_{11} = (0.04, 0.10, 0.18, 0.23; 0.9)$
	A_{12}	$W_{12} = (0.58, 0.63, 0.80, 0.86; 1.0)$	$S_{12} = (0.32, 0.41, 0.58, 0.65; 0.7)$
	A_{13}	$W_{13} = (0.0, 0.0, 0.02, 0.07; 1.0)$	$S_{13} = (0.58, 0.63, 0.80, 0.86; 0.8)$
C_2	A_{21}	$W_{21} = (0.04, 0.10, 0.18, 0.23; 1.0)$	$S_{21} = (0.93, 0.98, 1.0, 1.0; 0.85)$
	A_{22}	$W_{22} = (0.58, 0.63, 0.80, 0.86; 1.0)$	$S_{22} = (0.58, 0.63, 0.80, 0.86; 0.95)$
	A_{23}	$W_{23} = (0.0, 0.0, 0.02, 0.07; 1.0)$	$S_{23} = (0.32, 0.41, 0.58, 0.65; 0.9)$
C_3	A_{31}	$W_{31} = (0.04, 0.10, 0.18, 0.23; 1.0)$	$S_{31} = (0.17, 0.22, 0.36, 0.42; 0.95)$
	A_{32}	$W_{32} = (0.58, 0.63, 0.80, 0.86; 1.0)$	$S_{32} = (0.72, 0.78, 0.92, 0.97; 0.8)$
	A_{33}	$W_{33} = (0.0, 0.0, 0.02, 0.07; 1.0)$	$S_{33} = (0.58, 0.63, 0.80, 0.86; 1.0)$

Table 7.5 Evaluation on risk assessment for each company after aggregation

Company	Aggregated level of risk evaluation
C_1	(0.10, 017, 0.46, 0.71; 0.7)
C_2	(0.20, 0.30, 0.70, 1.00; 0.85)
C_3	(0.22, 0.31, 0.68, 0.98; 0.8)

Table 7.6 Evaluation on risk assessment for each company after aggregation

Company	Centroid-x	Centroid-y	Spread	Ranking value
C_1	0.4836	0.3601	0.2905	0.1236
C_2	0.4793	0.3445	0.1917	0.1326
C_3	0.3899	0.3549	0.2108	0.1092

7.7 Conclusion

In this chapter, a novel decision making methodology for grey number has successfully developed. This study first discussed the relevance of grey numbers in risk analysis decision making to ensure consistency of grey numbers with real world application. A special notion of grey numbers which is the capability to represent non-homogeneous data sets is pointed out in this study where two novel definitions of grey numbers value forms are given as part of the proposed work. Then, a novel consensus reaching method and ranking approach are proposed for the first time where both novelties have been validated as to demonstrate the novelty, validity and feasibility of the proposed work. Later on, this study exemplified the usefulness of the proposed work by applying the methodology developed towards a real world case study on risk assessment.

Although, it is acknowledged that fuzzy sets has received tremendous attentions from the practitioners and decision makers on its capability to resolve various decision problems, they fell short when it comes to deal with non-homogeneous data sets. In this case, grey numbers outperform fuzzy numbers in terms of dealing with non-homogeneous data sets efficiently. For future research, further investigations on computation of grey numbers need to be carried out as these efforts will support in addressing the incomplete and vague real-world information in a more flexible and accurate way.

References

1. Deng, J. (1989). Introduction to grey system theory. *Journal of Grey Systems, 1*(1), 1–24.
2. Lin, Y., Chen, M., & Liu, S. (2004). Theory of grey systems: Capturing uncertainties of grey information. *Kybernetes: The International Journal of Systems and Cybernetics, 33*, 196–218.
3. Liu, S., Gao, T., & Dang, Y. (2000). *Grey systems theory and its applications*. Beijing: The Science Press of China.
4. Liu, S., & Lin, Y. (2006). *Grey information theory and practical applications*. London: Springer.
5. Zadeh, L. (1965). Fuzzy sets. *Information and Control, 8*, 338–353.
6. Komorowski, J., Pawlak, J. Z., Polkowski, L., & Skowron, A. (1999). Rough sets: A tutorial. In S. K. Pal & A. Skowron (Eds.), *Rough fuzzy hybridization: A new trend in decision-making*. Singapore: Springer.
7. Yamaguchi, D., Li, G.-D., & Nagai, M. (2007). A grey-based rough approximation model for interval data processing. *Information Sciences, 177*, 4727–4744.
8. Wu, Q., & Liu, Z. (2009). Real formal concept analysis based on grey-rough set theory. *Knowledge-Based Systems, 22*, 38–45.
9. Deng, J. (1982). The control problems of grey systems. *Systems and Control Letters, 1*(5), 288–294.
10. Yang, Y., & John, R. (2012). Grey sets and greyness. *Information Sciences, 185*(1), 249–264.
11. Pham, V. N., Ngo, L. T., & Pedrycz, W. (2016). Interval-valued fuzzy set approach to fuzzy co-clustering for data classification. *Knowledge-Based Systems, 107*, 1–13.
12. Navara, M., & Navarová, M. (2017). Principles of inclusion and exclusion for interval-valued fuzzy sets and IF-sets. *Fuzzy Sets and Systems, 324*, 60–73.

13. Zadeh, L. A. (1975). The concept of a linguistic variable and its application to approximate reasoning-I. *Information Sciences, 8*, 199–249.
14. Wallsten, T. S., & Budescu, D. V. (1995). A review of human linguistic probability processing: General principles and empirical evidence. *The Knowledge Engineering Review, 10*(1), 43–62.
15. Yaakob, A. M., Serguieva, A., & Gegov, A. (2015). FN-TOPSIS: Fuzzy network for ranking traded equities. *IEEE Transaction on Fuzzy Systems, 25*(2), 315–332.
16. Deschrijver, G., & Kerre, E. (2003). On the relationship between some extensions of fuzzy set theory. *Fuzzy Sets and Systems, 133*(2), 227–235.
17. Haq, N., & Kannan, G. (2007). A hybrid normalised multi criteria decision making for the vendor selection in a supply chain model. *International Journal Management and Decision Making, 8*(5/6), 601–622.
18. Lin, Y. H., & Lee, P. C. (2007). Novel high-precision grey forecasting model. *Automation in Construction, 16*(6), 771–777.
19. Huang, S. J., Chiu, N. H., & Chen, L.W. (2008). Integration of the grey relational analysis with genetic algorithm for software effort estimation. *European Journal of Operational Research, 188*(3), 898–909.
20. Lin, Y. H., Lee, P. C., Chang, T. P., & Ting, H. I. (2008). Multi-attribute group decision making model under the condition of uncertain information. *Automation in Construction, 17*(6), 792–797.
21. Zavadskas, E. K., Kaklauskas, A., Turskis, Z., & Tamošaitiene, J. (2009). Multi-attribute decision making model by applying grey numbers. *Informatica, 20*(2), 305–320.
22. Van Laarhoven, P. J., & Pedrycz, W. (1983). A fuzzy extension of Saaty's priority theory. *Fuzzy Sets and Systems, 11*(1–3), 199–227.
23. Xu, L., & Wei, L. L. (2010). An improved method for ranking fuzzy numbers based on centroid. In *Seventh International Conference on Fuzzy Systems and Knowledge Discovery*, Yantai, Shandong (pp. 442–446).
24. Lin, J. G., Zhuang, Q. Y., & Huang, C. (2012). Fuzzy statistical analysis of multiple regression with crisp and fuzzy covariates and applications in analyzing economic. *Data of China Computational Economic, 39*, 29–49.
25. Bakar, A. S. A., & Gegov, A. (2014). Ranking of fuzzy numbers based on centroid point and spread. *Journal of Intelligent and Fuzzy System, 27*(3), 1179–1186.
26. Bakar, A. S. A., & Gegov, A. (2015). Multi-layer decision making methodology for ranking Z-numbers. *International Journal of Computational Intelligent Systems, 8*(2), 395–406.
27. Chutia, R. (2017). Ranking of fuzzy numbers by using value and angle in the epsilon-deviation degree method. *Applied Soft Computing, 60*, 706–721.
28. Alfonso, G., de Hierro, A. R. L., & Roldán, C. (2017). A fuzzy regression model based on finite fuzzy numbers and its application to real-world financial data. *Journal of Computational and Applied Mathematics, 318*, 47–58.
29. Aguilar-Peña, C., De Hierro, A. F. R. L., De Hierro, C. R. L., & Martínez-Moreno, J. (2016). A family of fuzzy distance measures of fuzzy numbers. *Soft Computing, 20*(1), 237–250.
30. Wang, X., Zhu, J., Song, Y., & Lei, L. (2016). Combination of unreliable evidence sources in intuitionistic fuzzy MCDM framework. *Knowledge-Based Systems, 97*, 24–39.
31. Bongo, M. F., & Ocampo, L. A. (2017). A hybrid fuzzy MCDM approach for mitigating airport congestion: A case in Ninoy Aquino International Airport. *Journal of Air Transport Management, 63*, 1–16.
32. Ghorabaee, M. K., Amiri, M., Zavadskas, E. K., & Antucheviciene, J. (2018). A new hybrid fuzzy MCDM approach for evaluation of construction equipment with sustainability considerations. *Archives of Civil and Mechanical Engineering, 18*(1), 32–49.
33. Shieh, B. S. (2007). An approach to centroids of fuzzy numbers. *International Journal of Fuzzy Systems, 9*, 51–54.
34. Wang, X., & Kerre, E. E. (2001). Reasonable properties for the ordering of fuzzy quantities (I). *Fuzzy Sets and Systems, 118*, 375–385.
35. Wang, X., & Kerre, E. E. (2001). Reasonable properties for the ordering of fuzzy quantities (II). *Fuzzy Sets and Systems, 118*, 387–405.

Chapter 8
Fuzzy Bayesian Nets and Influence Diagrams with Cognitive Numerical Judgment of Imprecise Probabilities

Aleksandar Janjić, Jelena D. Velimirović, and Lazar Z. Velimirović

8.1 Introduction

Organizations are required to do their business on the Internet if they want to stay competitive and survive on today's dynamic market. Economies and markets are becoming globalized due to affordable and user-friendly technology, and open Internet infrastructure. Even though electronic commerce or e-commerce usage in industry continues to grow worldwide due to no distance and time obstacles, lower transaction costs and more choices for customers [1], there are also some barriers in its adoption [2].

Online transactions and e-commerce increased organizations' effectiveness and improved their strategic positions, yet raised some new concerns related to trust, such as reliability of IT systems, low data security and privacy violation [3]. Customers who have doubts in e-services or interactions, will hardly ever engage in e-transactions or adopt e-commerce [4].

Perceiving e-commerce as unsafe and risky, emphasizes the importance of trust [5]. Therefore, customers' acceptance and usage of e-commerce is influenced by their own risk perceptions and trust which have been known for the most important psychological states that influence customers' online behavior [6]. We are aware that despite all of its advantages, the online setting is still characterized by uncertainty and fear of malicious behavior [7].

Therefore, in the context of digital economy, trust plays a central role. In [8] it is said that nowadays trust in commercial relationships is even more important

A. Janjić
Faculty of Electronic Engineering, University of Niš, Niš, Serbia
e-mail: Aleksandar.Janjic@elfak.ni.ac.rs

J. D. Velimirović · L. Z. Velimirović (✉)
Mathematical Institute of the Serbian Academy of Sciences and Arts, Belgrade, Serbia
e-mail: jelena.velimirovic@mi.sanu.ac.rs; lazar.velimirovic@mi.sanu.ac.rs

A. Meier et al. (eds.), *Applying Fuzzy Logic for the Digital Economy and Society*,
Fuzzy Management Methods, https://doi.org/10.1007/978-3-030-03368-2_8

than technology. It refers to reliability, partnership, and collaboration. Stable trust relationships are necessary for online business interactions. Without trust, online business relations will not be able to sustain.

In the context of digital economy, trust becomes the greatest company asset and its value can be seen in interactions between its community members. The real company value includes the platform/technology, but more importantly, it includes trust the company receives from its suppliers and buyers. Bayesian nets and influence diagrams facilitate the modeling of complex decision problems in digital economy, faced with many uncertainties. They are modeled using a compact graphical framework for representing the interrelationships between the variables involved in the problem under consideration. Both Bayesian net and Influence diagram, at the topological level represent an acyclic directed network with nodes representing variables critical to the problem and the arcs representing their interrelationships. Formal calculi have been developed for deterministic functions and probabilistic relationships based on either Bayesian or fuzzy probabilities.

Probabilistic uncertainties require appropriate mathematical modeling and quantification when predicting future state of the nature or the value of certain parameters. Imprecise probabilities can be expressed by fuzzy sets, linguistic terms or hybrid terms like: *most likely, improbable, fifty–fifty, around 30%*. These terms are in great extent influenced by the psychological profile of the decision maker. In this chapter, the biases when judging the values of low and high probabilities, together with the numerical size effect are modeled by psychophysics laws of stimulus response. Fuzzy probabilities are represented by triangular fuzzy numbers with lower and upper bounds determined by the quadratic programming optimization, with constraints on feasibility of elicited probabilities.

The chapter is organized as follows. In Sect. 8.2, the notion of trust in digital economy is explained. Section 8.3 gives the brief introduction to the possible biases Bayesian networks, and the notion of fuzzy probability. In Sect. 8.4, optimization of fuzzy probability elicitation based on cognitive judgment is presented, and finally, the optimization is illustrated in Sect. 8.5.

8.2 Trust in Digital Economy

8.2.1 Defining Trust

Everyday decision-making is based on trust—from what to eat and wear to solving complex problems. The notion of trust is inherent deeply in society due to its multi-dimensional nature and various contexts that it appears in. Therefore, trust is crucial for interactions on many levels/contexts—social, economic and political. It is highly subjective in nature and it relates to honesty, sincerity, truthfulness, reliability, dependability, confidentiality, etc., of the trusted person, consumer or agent [9].

When we trust in something or somebody we suppose that the result of our interaction will be positive for us; for example—we trust the goods we purchased from some website is of required quality, that the whole process of registering, ordering and paying for the goods is not risky or unsecure in any way, and that the bought items will be delivered as expected.

Trust arises with familiarity: if we already know something or it looks similar to something we already know, the positive outcome is expected and more likely to happen. This correlates to our interaction with friends, or with friends of friends, which is much easier than having to interact with complete strangers.

Therefore, the more transparent and trustworthy e-commerce is, the more customers will engage in it. This leads to familiarity, and wider adoption of e-commerce, thus economic growth [10]. The concept of trust has been studied across various disciplines such as economy, management, psychology, industrial engineering, computer science, etc. However, there is no consensus in literature on one universally accepted definition and there is no one way to measure trust value [6, 11]. The most adopted definition comes from the sociologist Diego Gambetta; he says that "Trust is a particular level of the subjective probability with which an agent assesses that another agent or group of agents will perform a particular action, both before he can monitor such action and in a context in which it affects its own action" [12].

Trust can also be defined as a consumer's expectation that an online interaction will go securely, without vulnerabilities exploited or any harm done to the consumer [13]. Some authors divide trust into general trust which refers to trust in IT infrastructure, and specific trust which is about trusting the website or platform a consumer wants to purchase from. Therefore, trust in the online setting can be defined as a consumer's confidence in the provider's reliability and benevolence to deliver the desired product or service on time, and the website or platform to be able to perform the supposed functions [14].

Trust can also be seen as belief of an agent, trustor, that the other agent, trustee, will act beneficially or positively. This means that trust is characterized by the actions performed by the trustee and their effect on the trustor's own actions [15].

8.2.2 The Importance of Digital Trust

Rapid ICT development transformed the world we live in and influenced all spheres of our lives—communication, health, education, work, banking, etc. Suddenly the world became hyper-connected with instant sharing of large amounts of data. Somehow our physical world got smaller, but the digital world became bigger with immense business opportunities. Economy transformed from its traditional model to a digital one. E-commerce is its first manifestation which enabled trade between companies (B2B), between companies and consumers (B2C) and recently between private individuals or peers (peer-to-peer: P2P).

Trust as well suffered transition from its offline to online version. This change offered us more choice to buy and sell globally and required new ways of building and receiving trust. It is now more than ever, that, in order to stay competitive, organizations need to be agile, creative, respond fast, and embrace all opportunities for digital growth. Companies/sellers should be able to secure their IT systems so that suppliers are confident the seller's IT infrastructure will not fail them, whereas customers wants to be sure their personal data is safe. When customers rely on the company's technology, and buy products from that company's website or platform, the company is considered trustworthy. If the purchase ends positively for the consumer, he will probably buy again from the same provider, thus become its loyal customer. This is one way of building trust between suppliers and sellers, or sellers and buyers. Apart from receiving trust from its suppliers and customers, the company enjoys good reputation as well.

Technology—resilient IT systems and secured and safe data—is important to create and encourage trust and build a good name for your company, but technology solely is not enough. Without consumers' participation and collaboration, generating digital trust will not be possible. Some ways to generate digital trust include creation of people's virtual identities which refers to more access to, and control, of their own personal and consumption data, transactions and operations; digitization of people's experience and opinions in form of rankings, ratings or comments left for a certain product or service; and creation of digital communities—for example, social media platforms people usually use to share knowledge and experience with their peers [16].

Buyers leave reviews online, usually on the company's website or platform, on products they bought, whereas potential buyers are able to see those reviews and ratings, and then form decisions on buying those products. In this case, decision-making is influenced by their peers' experience and recommendations. People usually rely on their peers' e-word of mouth [17]. Terms trust and reputation tend to be used interchangeably in the context of digital economy, even though they are not the same. Whereas trust is a subjective perception of an agent that other agent's actions will affect him positively, reputation includes what others think about you, or in the digital setting, what other people post about you, your company and brand [18]. This is how reputation economy is formed—new e-environment where companies are seen through lenses of other people's previous experiences.

8.2.3 Blockchain Technology and Trust

Nowadays e-commerce is susceptible to errors and needs certain improvements that include safe personal data and money transactions. Blockchain might be the key for these improvements. This new technology will influence industry-consumer relationships worldwide. To begin with, blockchain is a new technology that redefined the ways we buy and sell. In short, blockchain is a record of transactions, like a traditional ledger. Originally it was developed as part of digital currency Bitcoin.

Blockchain networks are decentralized, meaning transaction data are stored at many independent hard drives and servers around the world. This implies that it is almost impossible for hackers to control this network and violate stored data. Also, there isn't one central authority to change/validate information except those who are doing transactions. This technology eliminated the need for banks to act as a transaction third party which lead to no more transaction fees. It is a completely new way to transmit money from one person/organization to the next without using the traditional banking system. Due to its transparency and cost-effectiveness, blockchain offers more opportunity for building digital trust. Applications of blockchain technology are broad. Examples can be found mostly in the fields of the Internet of Things and financial services, especially areas and industries that work with payments, contracts (smart contracts), trust and others.

Even it is not perfect, and new technologies are emerging, blockchain networks have changed the nature of digital trust between organizations, customers, suppliers, and regulators. Companies that want to gain competitive advantage and improve their own financial performance will definitely need to invest in blockchain technology. The security and the trust in systems based on blockchain are established through cryptography and consensus algorithms (Proof-of-Work). Furthermore, in the specific area of smart contracts based on blockchain, another kind of trust is needed: the trust of programmers to the area specialists and lawyers. Therefore, the new models of subjective probabilities and trust have to be proposed. Bayesian networks with fuzzy probabilities as a practical tool to handle subjective probabilities in reasoning and decision making will be explained in the next section.

8.3 Bayesian Networks and Fuzzy Probabilities

8.3.1 Bayesian Network Biases

A large number of decision problems is facing the probabilistic uncertainty and imprecision when modeling problem structural parameters, including the required goals, constraints and external influences.

Bayesian networks and Influence diagrams are used as a convenient tool for the large class of these problems, while the inherent uncertainty has been modeled by the fuzzification of random variables, and/or prior and conditional probabilities. A comprehensive review of development dealing with imprecise probabilities for the solution of various engineering problems is given in [19]. Fuzzy probabilities are treated as an extension of interval probabilities, emphasizing the correspondence between different α-levels and probability boxes. Various engineering analysis are then enabled using min-max operator and extension principle as the basis for the processing of fuzzy information.

In Bayesian networks, uncertainty embodies both sources: aleatoric (random events or uncontrollable variation) and epistemic (as the absence of complete knowledge). Furthermore, fuzzy probabilities, grouped in several fuzzy sets, can be denoted with linguistic terms: extremely low, very low, medium, etc. [20, 21]. These terms represent the information granules that are in great extent influenced by the psychological profile of the decision maker.

Very few researchers dealing with interval or fuzzy probabilities assume that intervals or lower or upper bounds of fuzzy probability sets are not already known. In [22], the methods for obtaining interval probabilities based on pairwise comparison amongst all of possible outcomes has been proposed. Linear programming and quadratic programming methods are used to optimize interval entropy, variance and expected value. Although more advanced approach offers a formula to estimate interval probabilities from statistical viewpoint [23], the estimation of subjective probability intervals still represent a computational challenge.

For instance, if we imagine the numbers 0 through 100 (the subjective probability of certain states of nature) on a line, most people will say that they imagine a horizontal line, with 0 on the left, and an orderly progression to 100 on the right. However, the way of number progression depends on many different factors. The field of mathematical cognition (sometimes known as numerical cognition), reveals a fundamental characteristic of how numbers are represented in the brain, and it is well known that higher numbers are represented with lower fidelity than lower numbers. It has been shown that numerical processing, as with basic sensory modalities, obeys Weber's law such that the discriminability of two numbers decreases as the magnitude of the numbers increases [24], representing the so-called numerical size effect.

It is confirmed that logarithmic number line minimizes the error between input and representation relative to the subjective probability of number representation [25]. Generally, tendency exists to overestimate the position of relatively small numbers to the right. As a consequence, the positions of relatively large numbers are compressed toward the end of the scale [26]. In [27], a modified pairwise comparison procedure was used to empirically establish and assess membership functions for several probability terms. Results show lower range for terms like: *doubtful* and *improbable*, than *probably* and *likely*.

In decision making, [28] recognized that the human beings are subject to a number of biases distorting their judgment about the uncertainty of their knowledge. The most commonly biases are:

- the overestimation of the probability of events that are easy to recall,
- the focusing on possibly irrelevant characteristics of events in which they resemble other events,
- the bias of the final assessment of a value toward an initial assessment of the value by constraining subsequent adjustment of probabilities in the light of new evidence,
- overconfidence and underestimation of uncertainty about a quantity.

The usual explanation of the overestimation of low probabilities is that rare events tend to be overestimated because of the availability heuristic, anchoring on the "ignorance prior," and coarse chance categories. Furthermore, decision makers are more sensitive to probability changes close to 0 than to probability changes away from 0. The reason, often neglected in previous studies is that numerical judgments may be made at a different, less perceptual and more cognitive level. Another problem relative to subjective probabilities elicitation is the subjective inconsistency and non-additivity of mutually exclusive events.

Let us suppose that we have three mutually exclusive events, and the subjective assessment of their probabilities: *between 10% and 30%, between 20% and 40% and between 40% and 60%*. Intervals are corresponding to elicitators' judgment of *around 20%, around 30% and around 50%*. It is obvious that the sum of intervals *around 20% and around 30%* does not give the same interval as *around 50%*, violating the complementation property of probabilities. The similar problem of complementation, when both probability $P(A)$ and its complement $P(A_c)$ are right—crisp, left—fuzzy sets attaining 1 and 0.5 is given in [29].

8.4 Fuzzy Probabilities

The subjective probability can be granulated in different terms [20, 29, 30], but we will investigate the elicitation of triangular fuzzy set support—left and right bounds of triangular fuzzy numbers. Consider a discrete random variable X. We will assume that the probabilities of this random variables $FP(x_i)$ are assessed by a triangular fuzzy number:

$$\mu_{P_i}(X) = \begin{cases} 0, & x \le a_i \\ \frac{x-a_i}{b_i-c_i}, & a_i < x < b_i \\ \frac{c_i-x}{c_i-b_i}, & b_i < x < c_i \\ 0, & x \ge c_i \end{cases} \tag{8.1}$$

Let consider a set of fuzzy numbers $FP = \{FP_i = [a_i, b_i, c_i], i = 1, \ldots, n\}$. The interval of probability values for every α-cut will be denoted as $[a_{\alpha,i}, c_{\alpha,i}]$. We can interpret these fuzzy numbers as fuzzy probabilities as follows.

Definition 8.4.1 Fuzzy numbers $FP_i = [a_i, b_i, c_i]$ are called fuzzy probabilities of X if for $\forall\alpha \in [1, 0]$, and $\forall x_i \in [a_{\alpha,i}, c_{\alpha,i}]$ there are $x_i \in [a_{\alpha,i}, c_{\alpha,i}], \ldots, x_{i-1} \in [a_{\alpha,i-1}, c_{\alpha,i}]$, $x_{i+1} \in [a_{\alpha,i+1}, c_{\alpha,i+1}], \ldots, x_n \in [a_{\alpha,n}, c_{\alpha,n}]$ such that:

$$\sum_{i=1}^{n} x_i = 1 \tag{8.2}$$

Lemma 8.4.1 *The set of fuzzy numbers FP satisfies Eq.* (8.2) *if and only if the following conditions hold [31]:*

$$\begin{aligned} c_{\alpha,i} + a_{\alpha,1} + \ldots + a_{\alpha,i-1} + a_{\alpha,i+1} + \ldots + a_{\alpha,n} \leq 1, \forall\alpha, \forall i \\ a_{\alpha,i} + c_{\alpha,1} + \ldots + c_{\alpha,i-1} + c_{\alpha,i+1} + \ldots + c_{\alpha,n} \geq 1, \forall\alpha, \forall i \end{aligned} \tag{8.3}$$

Proof Sufficient conditions: If the first part of Lemma 8.4.1 holds, Eq. (8.3), then:

$$\begin{aligned} &\forall\alpha, \forall i\ x_i + a_{\alpha,1} + \ldots + a_{\alpha,i-1} + a_{\alpha,i+1} + \ldots + a_{\alpha,n} \leq c_{\alpha,1} + a_{\alpha,1} + \ldots + \\ &a_{\alpha,i-1} + a_{\alpha,i+1} + \ldots + a_{\alpha,n} \leq 1 \\ \\ &\forall\alpha, \forall i\ x_i + c_{\alpha,1} + \ldots + c_{\alpha,i-1} + c_{\alpha,i+1} + \ldots + c_{\alpha,n} \geq a_{\alpha,1} + c_{\alpha,1} + \ldots + \\ &c_{\alpha,i-1} + c_{\alpha,i+1} + \ldots + c_{\alpha,n} \geq 1 \end{aligned} \tag{8.4}$$

Then, the following expression holds:

$$\begin{aligned} &x_i + a_{\alpha,1} + \ldots + a_{\alpha,i-1} + a_{\alpha,i+1} + \ldots + a_{\alpha,n} \leq 1 \leq x_i + c_{\alpha,1} + \ldots + \\ &c_{\alpha,i-1} + c_{\alpha,i+1} + \ldots + c_{\alpha,n} \end{aligned} \tag{8.5}$$

The expression shows that there exist $a_{\alpha,j} \leq x_j \leq c_{\alpha,j}$, $j \in \{1, \ldots, n\} j \neq i$ that satisfies Eq. (8.2).

Necessary conditions. If the first part of Lemma 8.4.1 do not hold, Eq. (8.8), than:

$$\begin{aligned} \forall\alpha, \exists i\ c_{\alpha,i} + a_{\alpha,i} + \ldots + a_{\alpha,i-1} + a_{\alpha,i+1} + \ldots + a_{\alpha,n} > 1 \\ \forall\alpha, \exists i\ a_{\alpha,i} + c_{\alpha,i} + \ldots + c_{\alpha,i-1} + c_{\alpha,i+1} + \ldots + c_{\alpha,n} < 1 \end{aligned} \tag{8.6}$$

Then, taking x_i as $a_{\alpha,i}$ or $c_{\alpha,i}$ Eq. (8.2) cannot hold.

An alternative definition of fuzzy probabilities can be formulated from two extreme cases of $\alpha = 0$ and $\alpha = 1$.

Definition 8.4.2 Fuzzy numbers $FP_i = [a_i, b_i, c_i]$ are called fuzzy probabilities of X if for and $\forall x_i \in [a_i, c_i]$ there are $x_1 \in [a_1, c_1], \ldots, x_{i-1} \in [a_{i-1}, c_i] x_{i+1} \in [a_{i+1}, c_{i+1}], \ldots, x_n \in [a_n, c_n]$ such that:

$$\sum_{i=1}^{n} x_i = 1 \text{ and } \sum_{i=1}^{n} b_i = 1 \tag{8.7}$$

8.4.1 Bayesian Networks with Fuzzy Probabilities

Bayesian networks with fuzzy numbers replacing point value probabilities are proposed by Halliwell and Shen [29] defining "Bayesian fuzzy probability" as convex, normal fuzzy set of [0,1]. Complementation law has been relaxed in order to extend a partially defined linguistic probability measure, and this method has been successfully used in forensic statistics and risk analysis [21, 32]. More possible scenarios for fuzzifying the Bayesian approach are presented in [19] using non-fuzzy algorithmically efficient reformulation of the Bayesian formula.

The extension principle is used to define fuzzy counterparts to the standard arithmetic operators. The extension of a real arithmetic operator will be denoted by circling its usual symbol. It is also possible to derive these operators by examining the effects of interval based calculations at each α-cut. The extended operators are defined by Definition 8.4.3.

Definition 8.4.3 For all $a, b \in R^F$, the extended operators are defined by:

$$
\begin{aligned}
\mu_{A\oplus B}(z) &= \text{sum min}\begin{pmatrix} \mu_A(X), \ \mu_A(X) \\ x+y=z \end{pmatrix} \\
\mu_{A\otimes B}(z) &= \text{sum min}\begin{pmatrix} \mu_A(X), \ \mu_A(X) \\ xy=z \end{pmatrix} \\
\mu_{A-B}(z) &= \text{sum min}\begin{pmatrix} \mu_A(X), \ \mu_A(X) \\ x-y=z \end{pmatrix} \\
\mu_{A\oslash B}(z) &= \text{sum min}\begin{pmatrix} \mu_A(X), \ \mu_A(X) \\ x/y=z \end{pmatrix}
\end{aligned}
\tag{8.8}
$$

From previous definition, two fuzzy Bayes rules analogue to classical crisp number relations are formulated. Operator "$\cong$" stands for "$=$" operator.

Fuzzy joint probability:

$$P(Y = y_j, X = x_i) \cong P(X = x_i) \otimes P(Y = y_i \setminus X = X_i) \tag{8.9}$$

Fuzzy Bayes rule:

$$P(X = x_i \setminus Y = y_j) \cong \frac{P(X = x_i) \otimes P(Y = y_i \setminus X = X_i)}{P(Y = y_i)} \tag{8.10}$$

Based on the law of total probability another rule for the fuzzy marginalization can be added, represented by Eq. (8.11).

Fuzzy marginalization rule:

$$P(Y = y_j) \cong \sum_i P(X = x_i) \otimes P(Y = y_i \setminus X = X_i) \tag{8.11}$$

In the next section, we will give some introductory notes about the psychophysics, before defining the optimization problem of fuzzy probabilities adjustments.

8.5 Cognitive Numerical Judgment

In certain extent, the biases when judging the values of low and high probabilities are imposed by the psychological law of number representation. The numerical cognition reveals a fundamental characteristic of how numbers are represented in the brain. As stated before, the discriminability of two numbers decreases as the magnitude of the numbers increases. In this sense, the probability intervals can be treated as a perceived individual's response to the stimulus—the estimated probability magnitude. Weber's, Fechner's and Stevens' psychophysics law of stimulus response are most common models for the quantification of this relationship. Weber's and Fechner's law are based on "just noticeable difference threshold" concept.

8.5.1 Weber's Law

Weber's law claims that the noticeable change in perception is proportional to a relative increase of the stimulus. This kind of relationship can be described by a differential equation (8.12):

$$dp = k\frac{dS}{S} \tag{8.12}$$

where dp is the differential change in perception, dS is the differential increase in the stimulus and S is the stimulus at the instant. A constant factor k is to be determined experimentally. This "just noticeable difference" of the stimulus is always a constant percentage of the reference value. If the reference value is small, a small increase dp can be detected, and vice versa. Initially, this law was concentrated on perceptual values such as loudness or light, but [25] showed that more abstract parameters, including our sense of number, also followed Weber's law. Transposed to the probability judgment, the positions of relatively large numbers (analogue to stimulus S) are compressed toward the end of the scale biasing the uncertainty levels of small and large probabilities (fuzzy probability support).

8.5.2 Fechner's Law

Starting from the previous law, one can determine the relationship between the stimulus and response magnitudes. The larger the reference stimulus, the larger the increment necessary to give rise to a fixed change in perceived magnitude. This relationship is known as Fechner's law, obtained by the integration of Eq. (8.12), if we assume that both the internal and external domains are continuous spaces:

$$\begin{aligned} c &= k \ln S + C \\ C &= -k \ln S_0 \end{aligned} \tag{8.13}$$

where p is the perceived magnitude, k is a constant (Weber's Law), S is the physical magnitude of the stimulus parameter being investigated, S_0 is that threshold of stimulus below which it is not perceived at all. Equation (8.13) then becomes:

$$p = k \ln \frac{S}{S_0} \tag{8.14}$$

Equation (8.14) yield to the logarithmic relationship between stimulus and perception. The explanation for this effect in cognitive numerical judgment have been proposed by Lewicki and Bunker [3] arguing that the mental number line is logarithmically compressive, such that as numbers get larger they lie closer to each other. Therefore, discriminability decreases with increasing numerical magnitude because the distance between numbers becomes subjectively smaller as their magnitude increases.

The value corresponding to the $\mu(b_i)$ represents the stimulus level S. If we take the degree of uncertainty when determining the subjective probability (fuzzy probability support) for sensation level p, it is affected by the S level.

8.5.3 Stevens' Power Law

Fechner's law, Eq. (8.14), is a special case of Stevens' power law with three parameters. The power law encapsulates a broader range of possible data sets, varying the value of the exponent. For each different stimulus parameter there is some predictable relationship between the physical magnitude of that parameter and its perceived magnitude. Over a wide range this relationship can be expressed as the following function

$$p = k \cdot S^y \tag{8.15}$$

The exponent y is experimentally determined, and can be greater than one representing a positively accelerated or expansive function or it can be less than one representing a negatively accelerated or compressive function (approximating the Fechner's law which is compressive). Although the logarithmic scale seems to reflect better our internal representation of numbers (the normalization of industrial products such as bolt diameters or wheel sizes is based on logarithmic scale as well), the subjective probabilities bounds will be optimized according to both cases.

8.5.4 *Optimization Procedure*

In the case of the mental representation of number, instead of the linear model that allows for a simpler calculation of sums and differences, number cognition assumes a logarithmic encoding of number. According to Eq. (8.12), the ratio of fuzzy probability support $(c_i - a_i)$ and natural logarithm of stimulus intensity (b_i) will be constant (Eq. (8.16)).

$$\frac{c_i - a_i}{\ln(b_i)} = k_i \tag{8.16}$$

If we know in advance the vector of probability point values $b = [b_1, b_2, \ldots, b_n]$, the determination of fuzzy probability supports (a_i, c_i) is represented as the following minimization problem (Eq. (8.17)):

$$\begin{aligned} &\min_{a,c} F = \textstyle\sum_{i=1}^{n} (k_i - k_{avg})^2 \\ &k_{avg} = \frac{\sum\limits_{i=1}^{n} k_i}{n} \end{aligned} \tag{8.17}$$

s.t.

$$\begin{aligned} c_i + a_1 + \ldots + a_{i-1} + \ldots + a_n \geq 1, \forall i \\ a_i + c_1 + \ldots + c_{i-1} + \ldots + c_n \geq 1, \forall i \end{aligned} \tag{8.18}$$

An alternative way of supports optimization could be performed using Stevens' power law. In this case, the ratio of fuzzy probability support and stimulus intensity b_i^y will be constant:

$$\begin{aligned} &\min_{a,c} F = \textstyle\sum_{i=1}^{n} (k_i - k_{avg})^2 \\ &\tfrac{c_i - a_i}{b_i^y} = k_i \end{aligned} \tag{8.19}$$

s.t.

$$\begin{aligned} c_i + a_1 + \ldots + a_{i-1} + \ldots + a_n \geq 1, \forall i \\ a_i + c_1 + \ldots + c_{i-1} + \ldots + c_n \geq 1, \forall i \end{aligned} \tag{8.20}$$

The procedure for determination of optimal probability bounds, for both Fechner's and Stevens' equations will be explained on illustrative example. The optimization is performed using the interiorpoint approach to constrained minimization, solving a sequence of approximate minimization problems, with linear inequalities constraints.

8.6 Case Study

Illustrative example deals with the case very often in e-commerce transactions, concerning the order of certain goods via internet. Let suppose the situation where company has to make the decision of ordering 10,000 units of certain product. There are various suppliers with different prices and terms of delivery, that are not certain and we cannot be sure that the supplier will respect these conditions after the order has been done. However, we can use comments and impressions of existing clients, reflecting their subjective impression about the supplier credibility.

Prior beliefs about the supplier quality are given in Table 8.1, representing the state spaces of the variables with corresponding crisp probabilities (in the range from 0 to 1). We can rank the supplier credibility and quality by different grades, starting with three categories: *Problematic (P)*, *Reliable (R)*, and *Outstanding (O)*.

Ordering the product from the problematic supplier can cause delays and provoke costs to the company, but not buying can be treated as an opportunity cost, because the company may found the much better supplier for the same product. The utility function for different states of nature is given in Table 8.2, reflecting the buyer's satisfaction graded from 0 to 10.

Instead of ordering the whole quantity, the company may order the samples (100 units) to be sure about the supplier credibility and trust. The grades for the delivered products can be different, including different tests, and they will be denoted as *Bad (B)*, *Medium (M)*, and *Good (G)*. Based only on tests, we cannot be completely sure about the supplier's credibility. We can talk only on probability that the big order will be the same as the sample order. These conditional probabilities, estimated by experts are given in Table 8.3.

Table 8.1 Prior subjective probabilities

Supplier quality	Probability
Problematic	0.1
Reliable	0.3
Outstanding	0.6

Table 8.2 Utility function

	Buy the product	Do not buy
Problematic	0	6
Reliable	4	3
Outstanding	10	2

Table 8.3 Conditional probabilities of samples

	Quality of samples		
Quality	Bad	Medium	Good
Problematic	0.6	0.3	0.1
Reliable	0.3	0.4	0.3
Outstanding	0.1	0.4	0.5

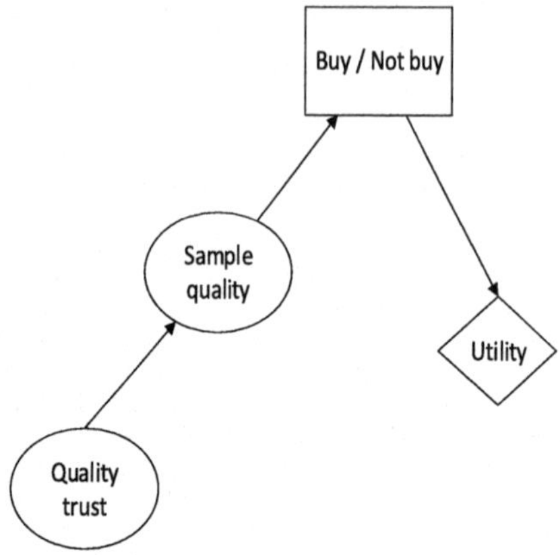

Fig. 8.1 The supplier selection problem

This problem can be represented by the Influence Diagram, with one decision node (Buy/Not buy), and two chance nodes: prior probability of supplier quality given in Quality trust node, and Sample quality node with results depending on supplier real quality. The Influence diagram is presented on Fig. 8.1.

Decision about the buying process is obtained comparing the expected utility for both cases. Let suppose that sample order was *Good* (*Sample quality* = *G*). Expected utility in both cases (*j* = *Buy, No buy*) is given in Eq. (8.21).

$$E(U)_j = \sum_{P,R,O} P(\text{Quality} = P, R, O/\text{Test} = G) \cdot U(j/P, R, O) \tag{8.21}$$

Let suppose that prior and conditional probabilities are obtained through the subjective experts' judgment. We will examine two variants of probability modeling, as two optimization problems: Fechner's and Stevens' law, Eqs. (8.14) and (8.15), with $y = 1$ as a parameter.

The optimization is performed using MatLab solver of quadratic function with linear constraints, and the results are presented in Tables 8.4 and 8.5.

It can be seen from previous tables, that logarithmic law gives more balanced results that Stevens law. In the latter case, probability supports are much lower at small than at larger probabilities. Using Eqs. (8.9)–(8.11), posterior probabilities of product quality are calculated and presented on Figs. 8.2 and 8.3. Expected utilities, calculated using Eq. (8.19) are presented on Figs. 8.4 and 8.5.

Subjective probability fuzzy numbers show the accordance with the empirically established membership functions for several probability terms [31]. Fechner's law

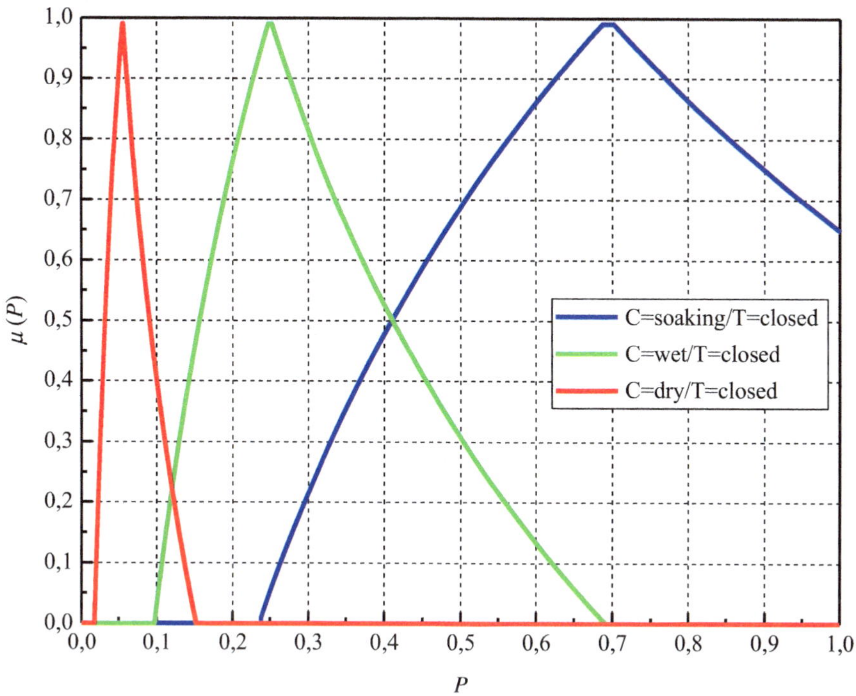

Fig. 8.2 Fuzzy posterior probabilities for Stevens's law

Table 8.4 Fuzzy probabilities for Fechner (F) and Stevens law (S)

	Test	
	S	F
	Bad	
Problematic	[0.48 0.6 0.72]	[0.49 0.6 0.72]
Reliable	[0.22 0.3 0.35]	[0.24 0.3 0.35]
Outstanding	[0.07 0.1 0.13]	[0.05 0.1 0.15]
	Medium	
Problematic	[0.22 0.3 0.34]	[0.19 0.3 0.38]
Reliable	[0.35 0.4 0.45]	[0.34 0.4 0.46]
Outstanding	[0.25 0.4 0.52]	[0.32 0.4 0.48]
	Good	
Problematic	[0.08 0.1 0.12]	[0.04 0.1 0.16]
Reliable	[0.22 0.3 0.35]	[0.24 0.3 0.35]
Outstanding	[0.34 0.5 0.65]	[0.42 0.5 0.59]

Table 8.5 Fuzzy prior probabilities

Prior	S	F	S	F	S	F
	[0.15 0.2 0.25]	[0.12 0.2 0.28]	[0.25 0.3 0.39]	[0.21 0.3 0.39]	[0.39 0.5 0.61]	[0.39 0.5 0.60]

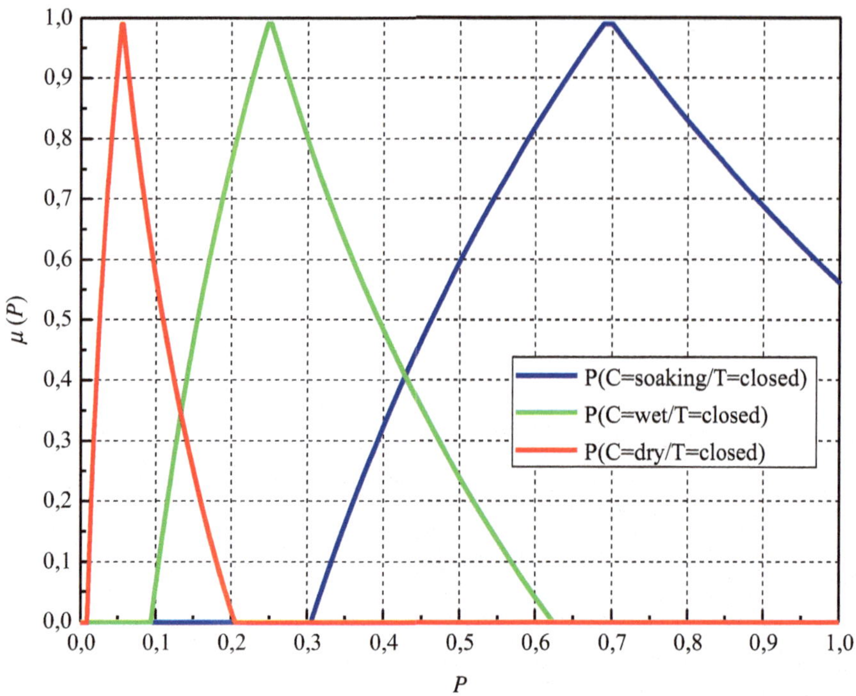

Fig. 8.3 Fuzzy posterior probabilities for Fechner's law

optimization results practically coincide with one of the subject's assessment of Improbable, Possible and Good Chance estimation.

Fuzzy prior probabilities and CPTs given in Tables 8.3 and 8.4 are practically simplified as crisp numbers given in Table 8.2. Therefore, the values corresponding to unity membership values on previous figures represent the results that would have been obtained using crisp values only. Graphical representations are sufficient to illustrate the possible error if we were to use only crisp probabilities, because of visible probability overlapping. The determination of appropriate exponent y in Stevens' law, and confirmation of Fechner's law should be the topic of further experiments and research concerning the risk attitude definition of decision maker. Our main concern was to model the biases inherent for humans' numerical judgment, especially in the space of probability assessment.

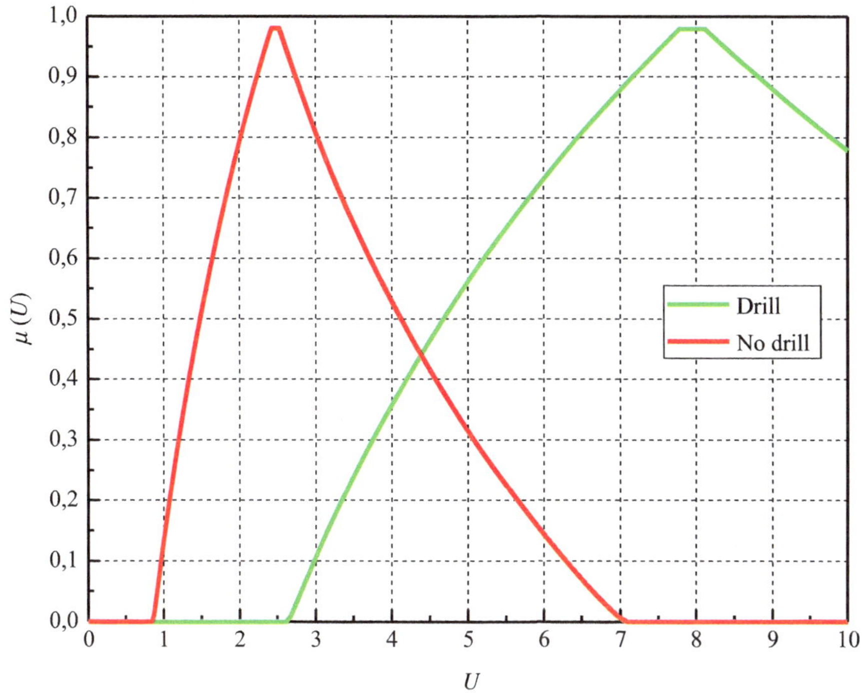

Fig. 8.4 Fuzzy expected utilities according to Stevens law

8.7 Conclusion

The cognitive science proved that the parameter that governs the ability to distinguish two numbers is not their absolute numerical distance, but distance relative to their size. Subjectively, the distance between 80 and 90 is not identical to that between 10 and 20. Furthermore, logarithmic number line is the one which minimizes the error between input and mental representation of a number. In this analysis, we established a connection between subjective uncertainty during the probability assessment and mental number representation of different probabilities expressed by approximate numbers.

Subjective probabilities, necessary for the modeling of trust in digital economy can be represented as information granules described by linguistic terms and modeled as triangular fuzzy numbers. The proposed optimization functions proved to be efficient in determination of feasible probability bounds, yet corresponding to the human cognitive process. The quadratic programming model is proposed that can be easily solved, and simulation results are concurrent with the experimental findings of subjective probability assessment.

Bayes networks and Influence diagrams can be used for wide range of tasks, including the reasoning and decision making under uncertainties. The proposed

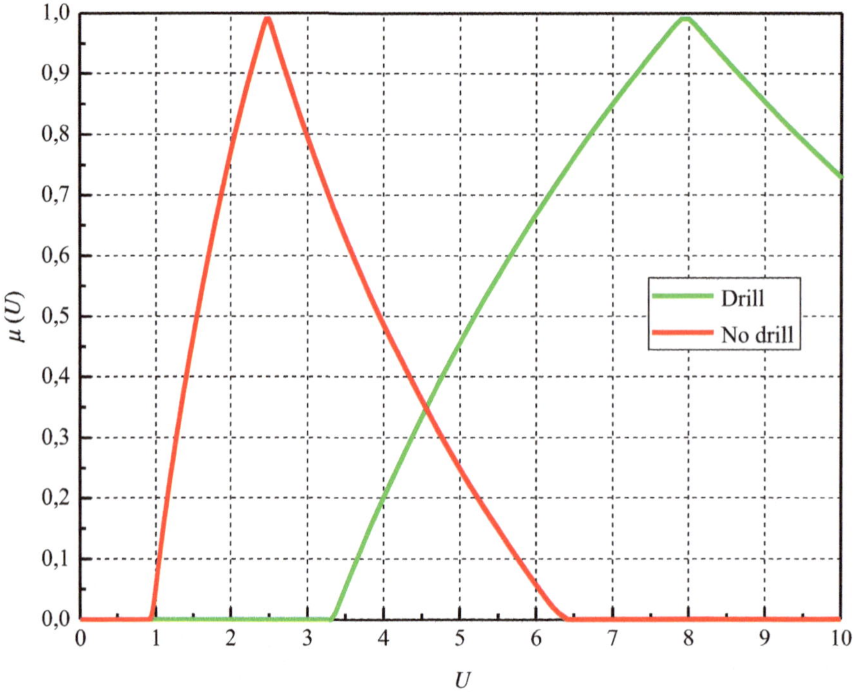

Fig. 8.5 Fuzzy expected utilities according to Fechner law

modeling of biases and human cognition in subjective probability elicitation makes their usage more comfortable in the problems of trust and human cognition process in the e-commerce. The model is illustrated on e-commerce decision making problem and obtained results showed that fuzzy probability model can be easily integrated into the existing decision making or risk assessment systems.

Acknowledgements This work was supported by the Ministry of Education, Science and Technological Development of the Republic of Serbia under Grant III 42006 and Grant III 44006.

References

1. Guo, J., & Sun, C. (2004). Global electronic markets and global traditional markets. *Electronic Markets, 14*(1), 4–12.
2. Chang, M. K., Cheung, W., & Tang, M. (2013). Building trust online: Interactions among trust building mechanisms. *Information & Management, 50*(7), 439–445.
3. Lewicki, R. J., & Bunker, B. B. (1996). Developing and maintaining trust in working relationships. In R. M. Kramer & T. Tyler (Eds.), *Trust in organizations* (pp. 114–139). Thousand Oaks: Sage.

4. Grabner-Kräuter, S., & Kaluscha, E. A. (2003). Empirical research in on-line trust: A review and critical assessment. *International Journal of Human-Computer Studies, 58*(6), 783–812.
5. Pavlou, P. A. (2003). Consumer acceptance of electronic commerce: Integrating trust and risk with the technology acceptance model. *International Journal of Electronic Commerce, 7*(3), 101–134.
6. Kim, D. J., Ferrin, D. L., & Rao, H. R. (2008). A trust-based consumer decision-making model in electronic commerce: The role of trust, perceived risk, and their antecedents. *Decision Support Systems, 44*(2), 544–564.
7. Head, M., & Yuan, Y. (2001). Privacy protection in electronic commerce–A theoretical framework. *Human Systems Management, 20*(2), 149–160.
8. Keen, P., Ballance, G., Chan, S., & Schrump, S. (1999). *Electronic commerce relationships: Trust by design*. Upper Saddle River: Prentice Hall PTR.
9. Yan, Z., & Holtmanns, S. (2008). *Trust modeling and management: From social trust to digital trust* (pp. 290–323). Hershey: IGI Global.
10. Jones, K., & Leonard, L. N. (2008). Trust in consumer-to-consumer electronic commerce. *Information & Management, 45*(2), 88–95.
11. Chang, W. L., Diaz, A. N., & Hung, P. C. (2015). Estimating trust value: A social network perspective. *Information Systems Frontiers, 17*(6), 1381–1400.
12. Gambetta, D. (2000). Can we trust trust. In *Trust: Making and breaking cooperative relations* (Chap. 13, pp. 213–237). Oxford: University of Oxford.
13. Beldad, A., De Jong, M., & Steehouder, M. (2010). How shall I trust the faceless and the intangible? A literature review on the antecedents of online trust. *Computers in Human Behavior, 26*(5), 857–869.
14. Thatcher, J. B., Carter, M., Li, X., & Rong, G. (2013). A classification and investigation of trustees in B-to-C e-commerce: General vs. specific trust. *Communications of the Association for Information Systems, 32*, 4.
15. Yau, S. S. (2006). Managing trust in distributed agent systems. In *International Conference on Autonomic and Trusted Computing* (pp. 17–25). Berlin: Springer.
16. Sundararajan, A. (2016). *The sharing economy: The end of employment and the rise of crowd-based capitalism*. Cambridge: MIT Press.
17. Lindgreen, A., & Vanhamme, J. (2005). Viral marketing: The use of surprise. In *Advances in electronic marketing* (pp. 122–138). Hershey: Group Publishing.
18. Ert, E., Fleischer, A., & Magen, N. (2016). Trust and reputation in the sharing economy: The role of personal photos in Airbnb. *Tourism Management, 55*, 62–73.
19. Beer, M., Ferson, S., & Kreinovich, V. (2013). Imprecise probabilities in engineering analyses. *Mechanical Systems and Signal Processing, 37*(1–2), 4–29.
20. Liu, J., López, L. M., Yang, J. B., & Wang, J. (2008). Linguistic assessment approach for hierarchical safety analysis and synthesis. In *Intelligent decision and policy making support systems* (pp. 211–230). Berlin: Springer.
21. Ren, J., Wang, J., Jenkinson, I., Xu, D. L., & Yang, J. B. (2007). A Bayesian network approach for offshore risk analysis through linguistic variables. *China Ocean Engineering, 21*(3), 371.
22. Guo, P., & Tanaka, H. (2010). Decision making with interval probabilities. *European Journal of Operational Research, 203*(2), 444–454.
23. Yager, R. R., & Kreinovich, V. (1999). Decision making under interval probabilities. *International Journal of Approximate Reasoning, 22*(3), 195–215.
24. Moyer, R. S., & Landauer, T. K. (1967). Time required for judgements of numerical inequality. *Nature, 215*(5109), 1519.
25. Dehaene, S. (2003). The neural basis of the Weber–Fechner law: A logarithmic mental number line. *Trends in Cognitive Sciences, 7*(4), 145–147.
26. Siegler, R. S., Fazio, L. K., Bailey, D. H., & Zhou, X. (2013). Fractions: The new frontier for theories of numerical development. *Trends in Cognitive Sciences, 17*(1), 13–19.
27. Wallsten, T. S., Budescu, D. V., Rapoport, A., Zwick, R., & Forsyth, B. (1986). Measuring the vague meanings of probability terms. *Journal of Experimental Psychology: General, 115*(4), 348.

28. Tversky, A., & Kahneman, D. (1974). Judgment under uncertainty: Heuristics and biases. *Science, 185*(4157), 1124–1131.
29. Halliwell, J., & Shen, Q. (2002). Towards a linguistic probability theory. In *Proceedings of the 2002 IEEE International Conference on Fuzzy Systems, 2002. FUZZ-IEEE'02* (Vol. 1, pp. 596–601). Honolulu: IEEE.
30. Janji, A., Stankovi, M., & Velimirovi, L. (2015). Multi-criteria influence diagrams – A tool for the sequential group risk assessment. In *Granular computing and decision-making* (pp. 165–193). Cham: Springer.
31. Weichselberger, K., & Pöhlmann, S. (1990). *A methodology for uncertainty in knowledge-based systems*. Heidelberg: Springer.
32. Halliwell, J., Keppens, J., & Shen, Q. (2003). Linguistic Bayesian networks for reasoning with subjective probabilities in forensic statistics. In *Proceedings of the 9th International Conference on Artificial Intelligence and Law* (pp. 42–50). New York: ACM.

Chapter 9
Oilfield Abandonment Decision by Applying a Fuzzy Pay-Off Method for Real Options

Roberto Evelim Penha Borges, Andreas Meier, Marco Antonio Guimarães Dias, and Adrião Duarte Dória Neto

9.1 Introduction

Petroleum exploration and production (E&P) is an activity that involves identifying potential oil and gas accumulations, drilling wells to extract the hydrocarbon and operating the whole structure for managing the oilfield. Dias [18] presents the typical E&P investment decisions phases, the last of which should be the oilfield abandonment. As stated by Parente et al. [29], this last stage highlights a difference from E&P to many others industries: the projects typically present an additional third period of cash flow—after the investment and production phases. This abandonment cash flow refers to all decommissioning expenses, which are costly and involves regulatory and environmental considerations [28]. By the end of production, besides the abandonment costs, companies should also account for the potential value of selling or reusing equipment. Therefore, there is a revenue that should be considered in the abandonment pay-off, making the decision and valuation more complicated.

The abandonment decision draws special attention when the rate of production of an oilfield approaches an economic limit below which continuing its production would result in a net loss. In principle, producers are supposed to abandon the field

R. E. P. Borges (✉) · A. D. D. Neto
Universidade Federal do Rio Grande do Norte, Natal, Brazil
e-mail: roberto.borges@petrobras.com.br; adriao@dca.ufrn.br

A. Meier
Université de Fribourg, Fribourg, Switzerland
e-mail: andreas.meier@unifr.ch

M. A. G. Dias
Pontifícia Universidade Católica do Rio de Janeiro, Rio de Janeiro, Brazil
e-mail: marcoagd@ele.puc-rio.br

A. Meier et al. (eds.), *Applying Fuzzy Logic for the Digital Economy and Society*, Fuzzy Management Methods, https://doi.org/10.1007/978-3-030-03368-2_9

as soon as its results are negative, that is when the revenues from the field are lower than the costs for producing its oil. However, the timing of abandoning is a tough decision because the uncertainty of the future increases the difficulty of ex-ante analysis [31]. Therefore, only the course of time will tell if the right call was made—both if the decision was to abandon or not. Assuming that the decommissioning is irreversible, the decision to abandon prunes all alternative development options and may avoid future profits, which might be possible under improving conditions. On the other hand, the company may have a difficult time with stakeholders if they continue an operation in conditions which cut into its profitability [9].

Following Dias [19], the traditional method to support the oilfield abandonment decision is to build yearly operational cash flow projections and suggest to produce until the year which has the last positive estimate, abandoning in the following year. In addition, as a way to consider the estimated abandonment cash flow into the analysis—especially when it is negative—some companies calculate the benefit of postponing this expense and producing even after the projection is negative. This approach, which may be seen as an opportunity cost analysis, basically accounts for the fact that investing the abandonment value in the financial market—instead of spending it to abandon—will yield some profit that may compensate the expected operational loss.

The aforementioned approaches have an important issue in common: they rely on one single mean estimate of cash flow items, for example: future production rate, petroleum price and abandonment cost. However, although the subsurface models used nowadays are very sophisticated, the future production rate remains uncertain [3]; several models try to represent the petroleum market behavior [18], but, as a commodity, its prices are unpredictable [25]; the abandonment costs are highly uncertain, mainly because of the industry's lack of experience [28].

Since our forecasting ability is limited, Bickel and Bratvold [3] suggest that the industry should focus on making good decisions, instead of reducing/removing uncertainty. In this way, less likely possibilities may carry important information regarding the decision [24], and therefore should be considered. Moreover, the static mean value used in the traditional methods assumes that the decision depends only on the readily available data, ignoring the additional information that might be revealed in the future.

The options offered by the described flexibility can be modeled by decision trees. However, Jafarizadeh and Bratvold [25] observe that the optimization that occurs at each downstream node changes the expected future cash flow of the project, which changes its risk characteristics and prevents the achievement of a correct result. The uncertain petroleum prices, complex cash flows structures, and interrelated decisions transform the timing of oilfield abandonment into a good example of complex real option [25].

After this introduction, the work follows with Sect. 9.2, that discusses real options valuation, citing its roots and objectives. Section 9.3 briefly describes fuzzy sets and presents their application in methodologies for real option valuation, including the one applied in this work: the Center of Gravity Fuzzy Pay-Off Method—CoG-FPOM. Section 9.4 comes up with a model for using the CoG-FPOM to support the oilfield abandonment decision, an example of the application of the model and its results. Section 9.5 finalizes the work with conclusions and suggestions for future works.

9.2 Real Options Valuation

In today increasingly complex world, uncertainty is present in most of the decisions that should be made by companies—including the oilfield abandonment decision, remarked in the introduction. Nevertheless, the traditional valuation methods typically utilize a single static mean value to support decisions, commonly using discounted cash flow (DCF) analysis and net present value (NPV) [24]. Besides having parameters difficult to estimate, those techniques do not consider less likely possibilities (potentially with high impact) in the analysis. In order to deal with the uncertainty—and the flexibilities—that this can offer to the decision makers, the real options analysis shows up as an important valuation tool.

Real options valuation is a methodology that highlights the value of managerial flexibility to respond optimally to the uncertainty. By observing that corporate investments opportunities can be viewed as financial call options on real assets, Myers coined in 1977 the term "real options" [18]. A real option is a right—not an obligation—to take an action on an underlying non-financial, real asset. The action may involve postponing a decision until a future time, abandoning, expanding or contracting a project, switching the input or the output, etc.

Tourinho developed the first real options mathematical model in 1979 [32]. Dixit and Pindyck published the first textbook in 1994 [20]. They pointed out the irreversibility, timing and uncertainty as key real options elements. The irreversibility (partial or total) increases the value of the "wait and see policy". The timing to exercise the option is then crucial to maximize the value of investment opportunity. The greater the uncertainty, the greater the value of flexibility, which is named the real options value when applicable to real assets investment. Dias [18] gives an overview of different real options models applied to petroleum assets.

Collan et al. [12] point out that real options analysis may be seen both as a qualitative method, like a mental model to analyze options for operational and strategic decision-making, and as a quantitative method, like a tool to perform numerical analysis for valuation purposes. The commonly used models for computing the real option value are based on the methods that have been used to value financial options: differential equation-based, especially Black-Scholes option pricing formula [4]; lattice-based, especially the binomial option valuation method [17]; and simulation-based methods, as the early example presented by Boyle [6].

Most of these models are complex and are based on the assumption that they can accurately mimic the underlying markets. This assumption may hold for some financial securities—like stocks and currencies, which are quite efficiently traded—but may not hold for real investments that do not have existing markets or whose markets don't exhibit even weak market efficiency [12]. An additional observation is that the traditional methods require the uncertainty to be typically of the parametric type, not considering structural or procedural uncertainty [13].

According to Favato et al. [23], real options research took the direction of searching for more sophisticated statistical models, increasing the complexity of calculus instead of focusing on management relevance. In the same direction, Mathews et al. [27] argue that the field of real options has been slow to develop because of the complexity of the techniques and the difficulty of fitting them to the realities of corporate strategic decision-making.

Favato et al. [23] are in favor of blending scenarios into real options valuation, arguing that companies should not be restricted to single forecasts, which are like predictions; instead, scenarios should be used as speculative descriptions of possible outcomes for the future, widening the chances of capturing potential opportunities and threats. By encouraging managers to envision future states of the world, scenario planning is a strategic management tool primarily used for qualitative analysis. If combined with real options, however, scenario planning may contribute to powerful quantitative assessments. In this way, decision-makers can work with a flexible valuation tool that is easy to understand and which can be lightly re-executed any time after the first decision is made—for example, when new information become available. This approach also allows for using separate risk adjusted discount rates for different cash flow items—like operational revenues, operational costs and capital investment—thus better representing the different types and levels of uncertainty within a project.

There are two main kinds of scenario-based methods for real options valuation: probability-based, like the Datar-Mathews method [27] and fuzzy-based, like the Fuzzy Pay-Off Method [12]. They both use forecasted projections for cash flows to derive a distribution of net present value for the project. Favato et al. [23] show that, all else equal, the application of a fuzzy-based method is feasible and useful without the necessity to engage in high-level and daunting mathematics. Nevertheless, Borges et al. [5] recently pointed out a technical inconsistency in the original Fuzzy Pay-Off Method and proposed a modified methodology for fuzzy real options valuation. The next section briefly describes fuzzy sets and presents their application in methodologies for real option valuation, including the original Fuzzy Pay-Off Method and the Center of Gravity Fuzzy Pay-Off Method.

9.3 Fuzzy-Based Real Options Valuation Methods

Zadeh [33] introduced fuzzy sets to mathematically represent imprecise and vague information and to provide formalized tools for dealing with these non-statistical uncertainties intrinsic to human language and perception—for example, in cash flow projections. Extending the classical crisp sets, to which an element may either belong or not, a fuzzy set assigns a real number between zero (complete non-membership) and one (complete membership) to each element of its universe of discourse—values in between represent a gradation of belonging. This flexibility may be helpful in making explicit the imprecision with which experts and modelers estimate parameter values used in models [13].

Let X be a nonempty classical set, known as the universe of discourse. A fuzzy set A of X is a mapping from X to the set [0, 1].[1]

$$A : X \to [0, 1] \tag{9.1}$$

The fuzzy set A is called normal if there exists an $x \in X$ such that $A(x) = 1$; otherwise it is called subnormal. The support of A is a crisp subset of X whose elements all have non-zero membership degrees in A: $supp(A) = \{x \in X | A(x) > 0\}$. The core of A is a crisp subset of X whose elements all have full membership degrees in A: $C(A) = \{x \in X | A(x) = 1\}$.

The support and the core may be seen as the largest and the smallest classical sets characterizing A, but sometimes it may be of interest to represent the fuzzy set by another crisp set between them. For $\alpha \in [0, 1]$, an α-level set (or α-cut) of A is defined by:

$$[A]^{\alpha} = x \in X | A(x) \geq \alpha \tag{9.2}$$

A fuzzy set A of X is called convex if $[A]^{\alpha}$ is a convex subset of X for all α (when $X = \mathbb{R}$, A is convex if $[A]^{\alpha}$ is a connected set, that is an interval, for all α). Finally, a fuzzy number A of X is a fuzzy set of the real line ($X = \mathbb{R}$) with a normal, convex and upper semi-continuous mapping function of bounded support [21].

Triangular fuzzy numbers are commonly used in problem modeling and may be seen as representing the statement "*x is approximately equal to a*". A triangular fuzzy number A with peak (or center) a, left width $\alpha > 0$ and right width $\beta > 0$ may be referenced as $A = (a, \alpha, \beta)$ and its mapping is defined by Eq. (9.3) and depicted by Fig. 9.1.

$$A(x) = \begin{cases} 1 - \frac{a-x}{\alpha}, & \text{if } a - \alpha \leq x \leq a \\ 1 - \frac{x-a}{\beta}, & \text{if } a \leq x \leq a + \beta \\ 0, & \text{otherwise} \end{cases} \tag{9.3}$$

Fuzzy numbers may be seen as possibility distributions [22, 34]. To notice the difference in the interpretation, consider a fuzzy number *young*, for which the numerical age $x = 28$ has a grade of membership $A_{young}(28) = 0.7$. The usual way of seeing this is that 0.7 depicts the degree of compatibility of 28 with the concept labeled young (fuzzy restriction). The other interpretation is that 0.7 represents the

[1]In this work we only consider type-1 fuzzy sets, for which the truth value algebra is the set [0, 1].

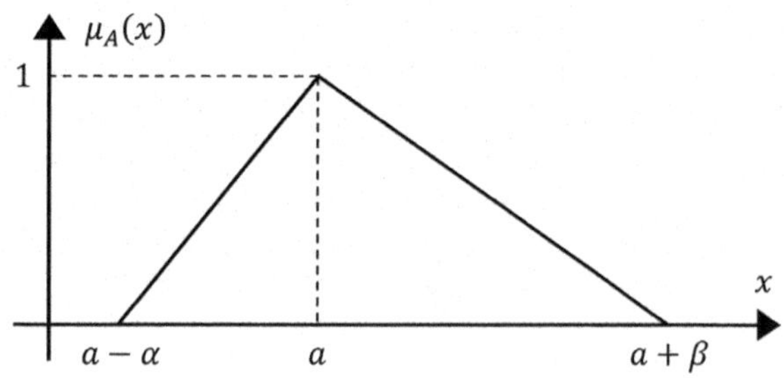

Fig. 9.1 Triangular fuzzy number

degree of possibility that somebody is 28 given the proposition that this person is young (possibility distribution).

In general, a variable may be associated both with a possibility distribution and a probability distribution, with the weak connection between the two expressed as the possibility/probability consistency principle [34]. Carlsson and Fullér [9] state that probability distributions can be interpreted as carriers of incomplete information, whereas possibility distributions can be interpreted as carriers of imprecise information. Kuchta [26] argues that probability theory is much less flexible than fuzzy sets theory because it has several assumptions about their distributions and operations that are seldom fulfilled in investment decisions cases. In practice, many times a company looks for a "good enough solution", which can be built by using fuzzy set theory [9]: at some point there will be a trade-off between precision and relevance, in the sense that increased precision can be gained only through loss of relevance and vice versa.

The literature on fuzzy real options is relatively recent: Carlsson and Fullér wrote one of the first papers in 2003 [8] and also published the first textbook on the subject in 2011 [9]. Collan et al. [13] recently made a survey regarding fuzzy numbers utilization in real options valuation. They show the use of fuzzy numbers together with differential equation-based models, lattice-based models and decision tree approaches. These fuzzy versions of real options analysis methods are generally usable under the same types of uncertainty as the underlying original methods with crisp numbers. Carlsson and Fullér [9] argue that a relevant reason for using fuzzy logic in real options valuation is that the imprecision encountered when judging or estimating future cash flows is not stochastic in nature, so that the use of probability theory may suggest a misleading level of precision and a notion that consequences are somehow repetitive.

9.3.1 The Center of Gravity Fuzzy Pay-Off Method (CoG-FPOM)

Fuzzy Pay-Off Methods use net present values of (usually three) cash flow scenarios to create a (usually triangular) pay-off fuzzy number—or possibility distribution. In

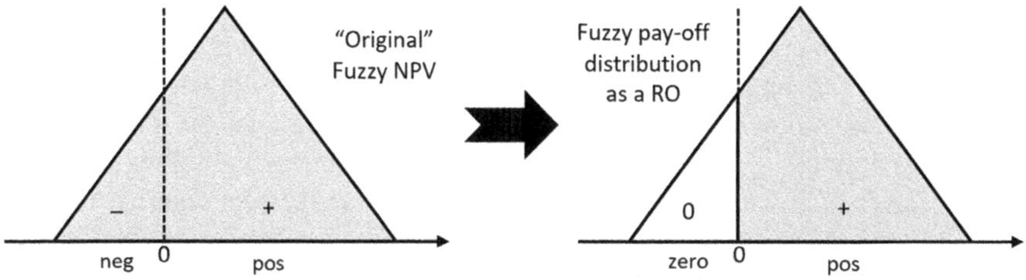

Fig. 9.2 FPOM's creation of pay-off distribution as real options, based on [13]

this way, that fuzzy number illustrates the degree to which a particular net present value estimate belongs to the set of possible net present values of the project. In order to include the real options flexibility within a project, the negative net present values of its possibility pay-off distribution are mapped into zero, reflecting the right of not proceeding with the project if a negative outcome is expected. Figure 9.2 illustrates this procedure, showing the "original" and the "modified" distributions.

In the sequence of the work, we will refer to the original fuzzy number as A (left side of Fig. 9.2) and to the modified fuzzy number, which has only the positive part of A as A_+ (right side of Fig. 9.2).

In order to obtain the value of the project with real options, it is necessary to calculate a most likely value of this modified distribution. In the original Fuzzy Pay-Off Method, this is done by calculating the possibilistic mean of the positive side of the distribution—according to the definition by Carlsson and Fullér [7]—and multiplying it by the fraction of the positive area of the distribution over its whole area. It is important to notice that this operation is effectively valuing all negative outcomes as zero. As defined by [12]:

$$ROV_{FPOM} = E(A_+) \times \frac{\int_0^\infty A(x)dx}{\int_{-\infty}^\infty A(x)dx} \tag{9.4}$$

where A stands for the fuzzy pay-off distribution; $E(A_+)$ denotes the possibilistic mean value of the positive side of A; $\int_0^\infty A(x)dx$ computes the area below the positive part of A and $\int_{-\infty}^\infty A(x)dx$ computes the area below the whole fuzzy pay-off distribution.

The Fuzzy Pay-Off Method for real options valuation has been used for analysis of research and development projects [16], patents [15], investments into information systems [11], corporate acquisitions [14], and large industrial investments [10]. As argued by Collan et al. [13], the method's input can range from hunches to detailed historical data-based information, which means that it can be useful not only under parametric, but also structural and procedural uncertainty. The price for this flexibility is that the output is not a precise real options valuation, but directions to be followed—which is in line with Bickel and Bratvold [3] reasoning. Finally, Favato et al. [23] show that although the Fuzzy Pay-Off Method simplifies the analysis, it offers sufficient precision in the results.

Recently a technical inconsistency was identified in the original Fuzzy Pay-Off Method [5]. As remarked in Sect. 9.2, real options analysis should add value to the company, either by upgrading profit opportunities or by mitigating downside risks. Ceteris paribus, a project with real options is worth more than the same project without real options—in the limit when the option is worthless, the values should be equal [1]. Even before the options pricing theory, the management science literature recognizes that "having the option to abandon never decreases project value; the typical consequences of ignoring the option would be to underestimate the value of a project" [30]. It happens that the original Fuzzy Pay-Off Method does not always follow this premise, and Borges et al. [5] identified situations in which the project without real options results in a higher value than the same project with real options.

In order to overcome this problem, Borges et al. [5] proposed the CoG-FPOM, a modified version of Fuzzy Pay-Off Method that uses the center of gravity (CoG) to make the approximation of a fuzzy number by a crisp number. According to this technique, the most representative value of the fuzzy number is the weighted average of the membership function [2] (Eq. (9.5)) and the value of a project with real options can then be computed using the CoG-FPOM (Eq. (9.6)).

$$CoG(A) = \frac{\int_{-\infty}^{\infty} xA(x)dx}{\int_{-\infty}^{\infty} A(x)dx} \tag{9.5}$$

$$ROV_{CoG\_FPOM} = CoG(A_{+}) \times \frac{\int_{0}^{\infty} A(x)dx}{\int_{-\infty}^{\infty} A(x)dx} \tag{9.6}$$

The CoG-FPOM was proved to have general validity, always delivering theoretically consistent results [5]. The calculation of $CoG(A_{+})$—the center of gravity of the positive side of the fuzzy pay-off distribution (see Fig. 9.2)—depends on where the zero pay-off is located within the fuzzy number. In order to have analytical solutions—which can be readily incorporated in spreadsheet software—Borges et al. [5] solved Eq. (9.5) for the four possible locations that the zero may be in relation to a triangular fuzzy number $A = (a, \alpha, \beta)$.

- Case 1: $0 < a - \alpha$

$$CoG(A_{+}) = \frac{3a - \alpha + \beta}{3} \tag{9.7}$$

In this situation, it is important to notice that the whole fuzzy number is above zero, and the center of gravity is calculated for the entire triangle (see Fig. 9.3). The result for this case is also used to calculate the ordinary center of gravity (Eq. (9.5)) for a triangular fuzzy number. The other 3 cases are represented by Eqs. 9.8, 9.9 and 9.10, and depicted by Figs. 9.4, 9.5 and 9.6 respectively.

- Case 2: $a - \alpha < 0 < a$

$$CoG(A_{+}) = \frac{\alpha(a+\beta)^3 - a^3(\alpha+\beta)}{3[\alpha(a+\beta)^2 - a^2(\alpha+\beta)]} \tag{9.8}$$

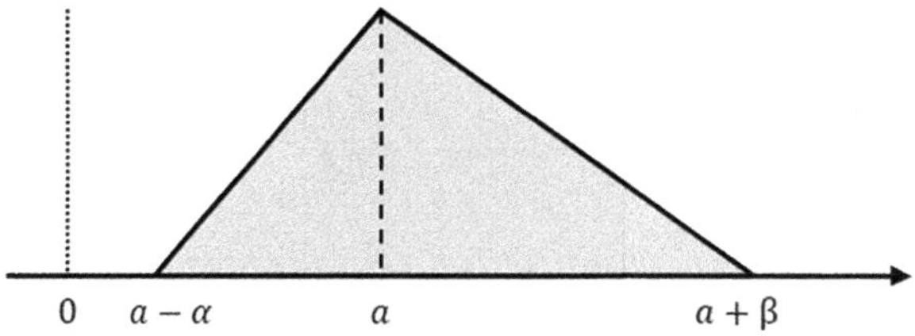

Fig. 9.3 Fuzzy pay-off distribution as real options with $0 < a - \alpha$ (from [5])

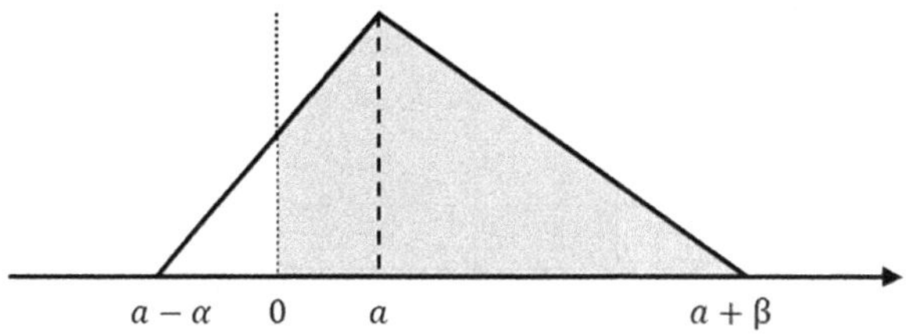

Fig. 9.4 Fuzzy pay-off distribution as real options with $a - \alpha < 0 < a$ (from [5])

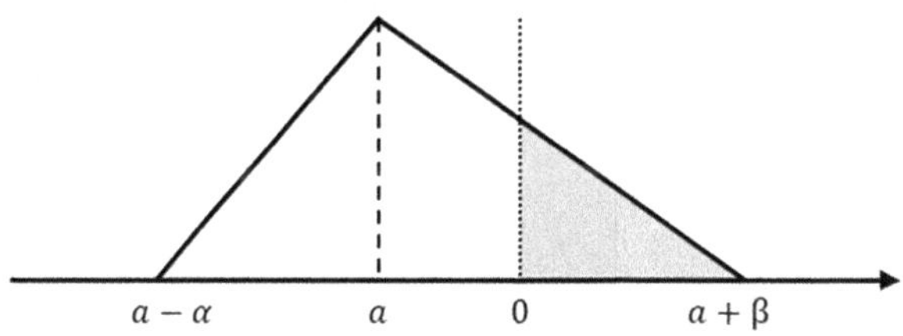

Fig. 9.5 Fuzzy pay-off distribution as real options with $a < 0 < a + \beta$ (from [5])

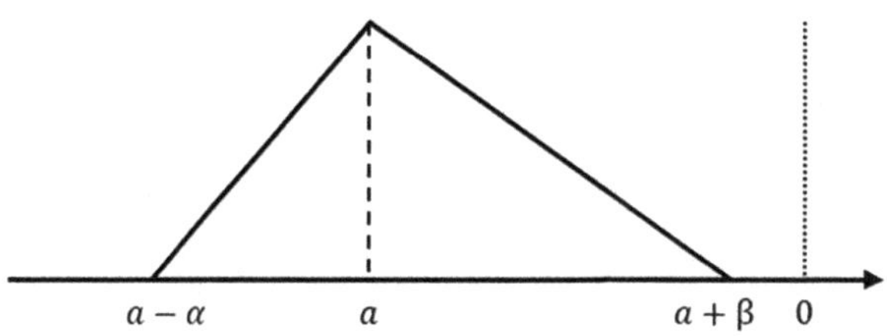

Fig. 9.6 Fuzzy pay-off distribution as real options with $a + \beta < 0$ (from [5])

- Case 3: $a < 0 < a + \beta$

$$CoG(A_+) = \frac{a + \beta}{3} \tag{9.9}$$

- Case 4: $a + \beta < 0$

$$CoG(A_+) = 0 \tag{9.10}$$

Based on the described CoG-FPOM method, the next section presents the proposed model for adequately valuing an oilfield and supporting its abandonment decision, and the application in an example together with its results and analysis.

9.4 A CoG-FPOM Model to Support Oilfield Abandonment Decision

In order to build the CoG-FPOM model to calculate the abandonment real option value for an oilfield, the starting point is the estimation of variables. Basically, the company compares the projected cash flow of keeping the production with the cash flow of abandoning it. Based on [19], the oilfield cash flows estimated for each year can be described as follows:

$$op_{CF} = prod \times price - (fixed_{cost} + var_{cost} \times prod) \tag{9.11}$$

$$ab_{CF} = res_{value} - ab_{cost} \tag{9.12}$$

where op_{CF} [MM US$] is the operating profit, $prod$ [MM un] is the petroleum production in that year, $price$ [US$/un] is the price of petroleum, already considering the benchmark crude oil projection and the spread to the specific oilfield petroleum price, $fixed_{cost}$ [MM US$] is the portion of the operating cost that does not depend on the rate of production, var_{cost} [US$/un] is the portion of the operating cost that depends on the rate of production, including the government take, ab_{CF} [MM US$] is the abandonment cash flow, res_{value} [MM US$] is the residual value of the oilfield and ab_{cost} [MM US$] is the abandonment cost, counting on environmental recovery. For the sake of simplicity, the income tax effect is not explicitly shown in Eqs. (9.11) and (9.12), but it does not change the qualitative results.

All the aforementioned variables are treated as uncertain and have their yearly values estimated/calculated for three scenarios[2]: an optimistic one, a most likely one and a pessimistic one. For example, in the optimistic scenario ab_{CF} is

[2]The presented model uses three scenarios and triangular fuzzy numbers, but it can be adapted to four scenarios and trapezoidal fuzzy numbers or whatever scenario strategy the company uses.

calculated to be the difference between the optimistically estimated res_{value} and the optimistically estimated ab_{cost}. It is important to notice that greater/lower values may be differently related to optimistic/pessimistic scenarios depending on the variable. For example, the optimistic res_{value} is greater than the pessimistic, since the result of the company is better in the case of a greater residual value. On the other hand, the pessimistic ab_{cost} is greater than the optimistic, since the result of the company is better in the case of a lower abandonment cost.

From the three ab_{CF} estimates for each year, it is possible to calculate one expected abandonment pay-off using the center of gravity (see Eqs. (9.5) and (9.7)). It is important to notice that the expected abandonment pay-off has to be estimated up to 1 year after the final year of forecasted production. This is because the end of the last year is the expiration of the option to produce and the company has no choice: the field has to be abandoned.[3] It means that the expected pay-off for the year following the end of production is its expected abandonment pay-off.

In order to achieve a result, the model follows a backwards decision strategy. For didactic purposes, let's consider that the last year that has production in the forecast is year 10. In the beginning of year 10, the company would have to decide between stopping or keeping the production. Following the proposed model, decision-makers would behave rationally and seek the real options value related to this flexibility. The three op_{CF} estimates for year 10, together with the expected pay-off for year 11, make it possible to build the corresponding fuzzy number.[4] The expected abandonment pay-off of year 10 defines the threshold below which the projections should be valued as zero, making it possible to use one of the 4 cases derived from Eq. (9.5). This calculated real option value becomes the expected pay-off for year 10 in case the company decides to produce that far. Following the backwards process, it is possible to calculate the estimated value of the field with real options at present.

Algorithms 1 and 2, presented below, intend to summarize the steps described above. Every underlined variable is an array of 3 floats representing a triangular fuzzy number of the form $A = (a; \alpha; \beta)$. It is important to notice that $a = A_{base}$; $\alpha = A_{base} - A_{pess}$; and $\beta = A_{opti} - A_{base}$ (see Fig. 9.1).

The value of the real option of abandoning the oilfield is calculated from the difference between the value of the field with real options (Algorithm 1) and the value of the field without real options. This last element can be calculated by applying Eq. (9.7) to the triangular fuzzy numbers of each year, without disregarding its negative side—which is similar to ignoring the integral terms in Eq. (9.6). After discounting and summing the elements, the value of the field without the option is calculated, and therefore the real options value can be achieved.

[3]The most common reasons are technical (life of equipment/facilities) or contractual (end of concession period).

[4]The expected pay-off for the following year must always be discounted to the year of the analysis—in our example, the expected pay-off for year 11 must be discounted to year 10.

Algorithm 1 Value of the oilfield with abandonment real option

```
n : integer ← quantity of years with projection
expPO : float[n + 1]                                ▷ expected pay-off for each year
expPO(n + 1) = CoG(abCF_{n+1})   ▷ for year n + 1, it is the abandonment pay-off (footnote 2)
for i = n to 0 do
    cont_i = opCF_i + DISCOUNT(expPO(i + 1))        ▷ pay-off for continuing the production
    APO_i = CoG(abCF_i)                             ▷ pay-off for abandoning in year i
    expPO(i) = CoG*(cont_i, APO_i) × A*(cont_i, APO_i)   ▷ ROV using CoG-FPOM
end for
oilfieldValue = expPO(0)
```

Algorithm 2 Functions used in the oilfield value calculation

```
function CoG(x)                      ▷ ordinary center of gravity of x (Eq. (9.7))
    return (3x_a − x_α + x_β)/3
end function

function DISCOUNT(PO)                ▷ discounts PO back one year using rate r
    return PO/(1 + r)
end function

function COG*(x, t)                  ▷ center of gravity of x disregarding values less than t
    if t < x_a − x_α then
        return (3x_a − x_α + x_β)/3                                       ▷ Eq. (9.7)
    else if t < x_a then
        return [x_α(x_a + x_β)^3 − x_a^3(x_α + x_β)]/{3[x_α(x_a + x_β)^2 − x_a^2(x_α + x_β)]}   ▷ Eq. (9.8)
    else if t < x_a + x_β then
        return (x_a + x_β)/3                                              ▷ Eq. (9.9)
    else
        return 0                                                          ▷ Eq. (9.10)
        fracA = 0
    end if
end function

function A*(x, t)                    ▷ fraction of area under x greater than t
    totalArea : float ← area under x
    prodArea : float ← area under x greater than t
    return prodArea ÷ totalArea
end function
```

9.4.1 Application of the Proposed Model

As highlighted in Sect. 9.2, it is common practice in companies to work with scenarios, which are carefully built and justified by strategy teams. In this application the projections were made by the authors, as described below, and are synthetic data.

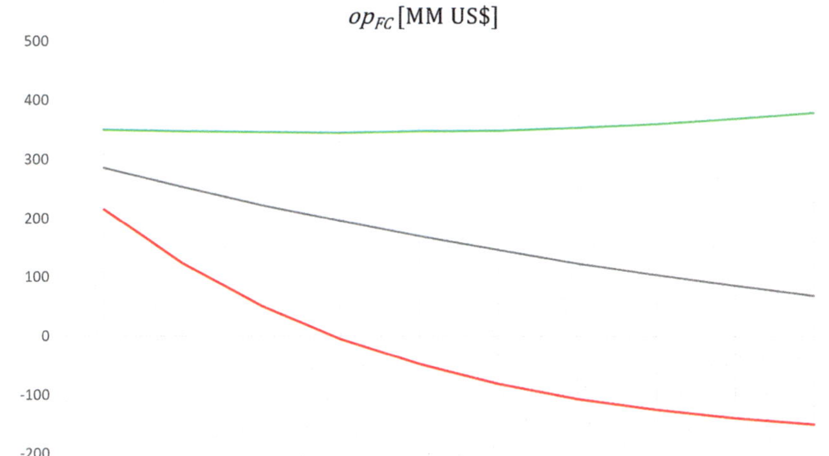

	1	2	3	4	5	6	7	8	9	10
Pess	215	123	51	-5	-48	-81	-107	-125	-139	-149
Base	285	253	223	195	170	146	124	104	86	69
Opti	350	347	346	345	347	349	354	360	369	380

Fig. 9.7 Forecasts of op_{CF} for application of the proposed model

Nevertheless, the example has been worked out with experienced managers[5] and can be viewed as representing well worked out scenarios.

For the sake of simplicity but without loss of generality, this application has some simplifications. The estimated production $prod$ is considered to start from an reasonable initial value and follow an exponential decline—a well-known analytical technique for petroleum production forecasting. Furthermore, res_{value} is considered to fall linearly from an initial value to zero, reflecting the wear and tear of the facilities. Finally, this application does not include royalty (whose effect is only price reduction) nor income tax (that does not change the cash flow signal) in var_{cost}.

Obviously, the same framework could be adapted to a producer's real projection of variables, both production related (like reservoir simulation outputs) and market related (like guesses for the world economy, which influences the oil price, exchange rates, etc.). Also, if any of the variables is not included in scenario planning or technical estimations, the analyst may use its most likely value directly into the model.

The hypothetical oilfield of this example has an initial oil production rate of 3000 m^3/d and an initial production cost of US$ 26.68 per barrel. Charts of the calculated op_{CF} and ab_{CF} used in this example are shown in Figs. 9.7 and 9.8, respectively.

[5]Managers of the petroleum exploration and production industry who have at least 10 years of experience in the planning/management/projection activity.

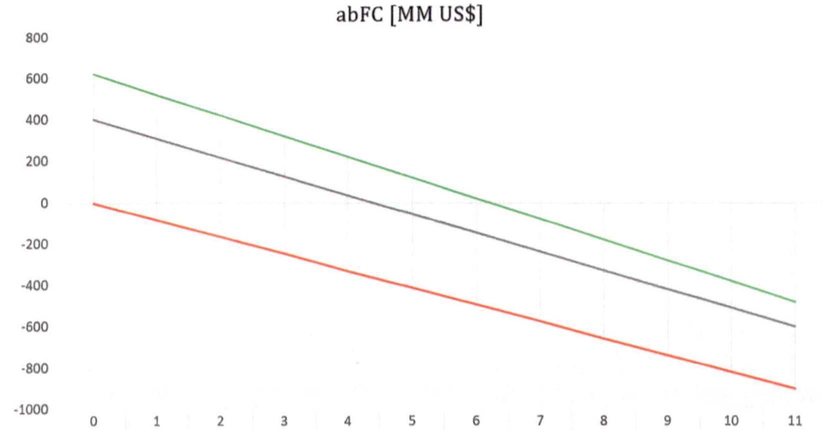

	0	1	2	3	4	5	6	7	8	9	10	11
Pess	0	-82	-164	-245	-327	-409	-491	-573	-655	-736	-818	-900
Base	400	309	218	127	36	-55	-145	-236	-327	-418	-509	-600
Opti	620	520	420	320	220	120	20	-80	-180	-280	-380	-480

Fig. 9.8 Forecasts of ab_{CF} for application of the proposed model

For this example, after running all the calculations and using a single discount rate of 10% per year, the abandonment real option value was 159 MM US$, obtained from the difference between the value of the oilfield with real option (1123 MM US$) and the value of the oilfield without real option 964 MM US$). This positive result indicates that the possibility of being able to abandon increases the value of the field, as expected. It also shows numerically what is an estimated value of this increase: approximately 16%. In this example, both values of the field were positive—meaning the company should decide to keep producing even if not considering the option. Nevertheless, in some specific circumstances the value of the field without real options may be negative while the value of the field with real options is positive. In those cases, the presented model would suggest to keep producing whereas the traditional methods would suggest to abandon.

9.5 Conclusions

This work presented a fuzzy scenario-based model for valuing the abandonment real option of an oilfield. The model is based in a fuzzy method with general validity, meaning that its results will always be consistent with real options theory and general management intuition.

An application was made in a hypothetical petroleum field. The results show how the model allows users to calculate the value of the oilfield with/without real options and consequently the value of the option itself. The example also showed the usefulness of the proposal in supporting the tough business decision of abandoning or not an oilfield.

One point to be further considered in the current model is that it simplifies the possibilities of values in between the scenarios by one straight line. Even assuming that it is not possible to perfectly model this transition, it might be interesting to study the variables and use a different shape for each of the possibilities distributions.

It is also possible to enhance the model by adding an expected abandonment year projection, because many times the decision is not to either stop now or produce until the end. This estimated year is very important in practice for corporate planning purposes. It also has impacts in reserves estimation, which influences other subjects, like impairment tests and depletion rate of assets.

Acknowledgements The authors would like to acknowledge the great technical support received from Mikael Collan, Roberto Teodoro and from the members of the Research Center for Fuzzy Management Methods of the University of Fribourg. Roberto E. P. Borges acknowledges the financial support received from Capes/Brazil through the PDSE scholarship number 88881.131650/2016-01. Marco A. G. Dias acknowledges the financial support from Petrobras/Cenpes through the research project FlexWell.

References

1. Amram, M., & Kulatilaka, N. (1999). *Real options: Managing strategic investment in an uncertain world*. Boston: Harvard Business School Press.
2. Bai, Y., & Wang, D. (2006) *Fundamentals of fuzzy logic control – Fuzzy sets, fuzzy rules and defuzzifications* (pp. 17–36). London: Springer London.
3. Bickel, J. E., & Bratvold, R. B. (2008). From uncertainty quantification to decision making in the oil and gas industry. *Energy Exploration & Exploitation, 26*(5), 311–325.
4. Black, F., & Scholes, M. (1973). The pricing of options and corporate liabilities. *Journal of Political Economy, 81*(3), 637–654.
5. Borges, R. E. P., Dias, M. A. G., Neto, A. D. D., & Meier, A. (2018). Fuzzy pay-off method for real options: The center of gravity approach with application in oilfield abandonment. *Fuzzy Sets and Systems, 353*, 111–123.
6. Boyle, P. P. (1977). Options: A Monte Carlo approach. *Journal of Financial Economics, 4*(3), 323–338.
7. Carlsson, C., & Fullér, R. (2001) On possibilistic mean value and variance of fuzzy numbers. *Fuzzy Sets and Systems, 122*(2), 315–326.
8. Carlsson, C., & Fuller, R. (2003). A fuzzy approach to real option valuation. *Fuzzy Sets and Systems, 139*(2003), 297–312.
9. Carlsson, C., & Fullér, R. (2011). *Possibility for decision: A possibilistic approach to real life decisions*. Berlin: Springer Publishing Company.
10. Collan, M. (2011). Valuation of industrial giga-investments: Theory and practice. *Fuzzy Economic Review, XVI*(1), 21–37.
11. Collan, M., Björk, K.-M., & Kyläheiko, K. (2014). Evaluation of an information systems investment into reducing the bullwhip effect – A three-step process. *International Journal of Logistics Systems and Management, 17*(3), 340–356.
12. Collan, M., Fullér, R., & Mezei, J. (2009). A fuzzy pay-off method for real option valuation. *Journal of Applied Mathematics and Decision Sciences, 2009*, 238196.

13. Collan, M., Haahtela, T., & Kyläheiko, K. (2016). On the usability of real option valuation model types under different types of uncertainty. *International Journal of Business Innovation and Research, 11*(1), 18.
14. Collan, M., & Kinnunen, J. (2011). A procedure for the rapid pre-acquisition screening of target companies using the pay-off method for real option valuation. *Journal of Real Options and Strategy, 4*(1), 117–141.
15. Collan, M., & Kyläheiko, K. (2013). Forward-looking valuation of strategic patent portfolios under structural uncertainty. *Journal of Intellectual Properties Rights, 18*, 230–241.
16. Collan, M., & Luukka, P. (2014). Evaluating R&D projects as investments by using an overall ranking from four new fuzzy similarity measure-based TOPSIS variants. *IEEE Transactions on Fuzzy Systems, 22*(3), 505–515.
17. Cox, J., Ross, S., & Rubinstein, M. (1979). Option pricing: A simplified approach. *Journal of Financial Economics, 7*(3), 229–263.
18. Dias, M. A. G. (2004). Valuation of exploration and production assets: An overview of real options models. *Journal of Petroleum Science and Engineering, 44*(1), 93–114.
19. Dias, M. A. G. (2014). *Análise de Investimentos com Opções Reais*. Rio de Janeiro: Interciência.
20. Dixit, A. K., & Pindyck, R. S. (1994) *Investment under uncertainty*. Princeton: Princeton University Press.
21. Dubois, D. (1997). *Fuzzy sets and systems: Theory and applications*. Orlando: Academic.
22. Dubois, D., & Prade, H. (1997). The three semantics of fuzzy sets. *Fuzzy Sets and Systems, 90*(2), 141–150.
23. Favato, G., Cottingham, J. A., & Isachenkova, N. (2015). Blending scenarios into real options: Relevance of the pay-off method to management investment decisions. *Journal of Finance and Accounting, 3*(2), 12–17.
24. Ho, S. H., & Liao, S. H. (2011). A fuzzy real option approach for investment project valuation. *Expert Systems with Applications, 38*(12), 15296–15302.
25. Jafarizadeh, B., & Bratvold, R. (2012). Two-factor oil-price model and real option valuation: An example of oilfield abandonment. *SPE Economics and Management, 4*(3), 158–170.
26. Kuchta, D. (2000). Fuzzy capital budgeting. *Fuzzy Sets and Systems, 111*(3), 367–385.
27. Mathews, S., Datar, V., & Johnson, B. (2007). A practical method for valuing real options: The Boeing approach. *Journal of Applied Corporate Finance, 19*(2), 95–104.
28. Osmundsen, P., & Tveteras, R. (2003). Decommissioning of petroleum installations—Major policy issues. *Energy Policy, 31*(970911), 1579–1588.
29. Parente, V., Ferreira, D., Moutinho dos Santos, E., & Luczynski, E. (2006). Offshore decommissioning issues: Deductibility and transferability. *Energy Policy, 34*(15), 1992–2001.
30. Robichek, A. A., & Horne, J. C. (1967). Abandonment value and capital budgeting. *Journal of Finance, 22*(4), 577–589.
31. Taleb, N. N. (2007). Black swans and the domains of statistics. *The American Statistician, 61*(3), 198–200.
32. Tourinho, O. A. F. (1979) *The valuation of reserves of natural resources—An option pricing approach*. PhD thesis. Berkeley: University of California.
33. Zadeh, L. A. (1965). Fuzzy sets. *Infection Control, 8*(3), 338–353.
34. Zadeh, L. A. (1978). Fuzzy sets as a basis for a theory of possibility. *Fuzzy Sets and Systems, 1*(1), 3–28.

Chapter 10
A Fuzzy-Based Recommender System: Case Börsenspiel for Swiss Universities

José Mancera, Minh Tue Nguyen, and Edy Portmann

10.1 Introduction

The stock market is one of the most dynamic environments, where many parameters change almost every second in order to provide the most updated data of the stocks. As a consequence the understanding and rapid interpretations of the information can make a significant impact on the revenue among different investment choices.

Analytics has been a tool for finance specialists or investors to plan their investment portfolios. Although analytics provide a good base to help investors to select stocks, the reality is that most financial analysts base their decisions on intuition and practical experience rather than on analytics. Understanding the main investment factors and integrating them into an algorithm that can recommend stocks to invest creates several research questions such as: what is the impact of the recommendations? How many investment strategies are needed? What is the algorithm performance? Who takes better investment decisions in rounds of investments, the recommender system or a investor?

The present study seeks to merge fuzzy logic with a recommender system model by creating a customized fuzzy-recommender algorithm, which can be applied in a

The current chapter is based on the work of José Mancera and Minh Tue Nguyen in the Web Analytics & Monitoring seminar research: "Fuzzy Recommender Systems in the Stock Market" at University of Fribourg, Switzerland, May 2015.

J. Mancera (✉)
Department of Informatics, University of Fribourg, Fribourg, Switzerland
e-mail: josealberto.manceraandrade@unifr.ch

M. T. Nguyen · E. Portmann
Human-IST Institute, University of Fribourg, Fribourg, Switzerland
e-mail: minhtue.nguyen@unifr.ch; edy.portmann@unifr.ch

A. Meier et al. (eds.), *Applying Fuzzy Logic for the Digital Economy and Society*, Fuzzy Management Methods, https://doi.org/10.1007/978-3-030-03368-2_10

real trading platform, such as the Börsenspiel for Swiss Universities (BSU) which is used as a case study in this chapter. It is a real simulator for the stock market where users can register and start to invest in order to practice their investment skills. The design and implementation of this recommender system (RS) can help to understand the impacts of fuzzy logic in terms of trading performance in comparison with amateurs and professional traders. In addition, this RS also provides the understanding of the stock characteristics, investment context, alternatives and its fuzzification.

The methodology is based on existing literature on business analytics, recommender systems, fuzzy mathematics, and also on the concrete BSU website. These sources will allow us to identify the components that play important roles in the design of a fuzzy recommender algorithm from the business analytics perspective. The output of the study contains, on the one hand, a presentation of the design, and the implementation of the FRS. On the other hand, a comparative discussion between analytics and intuition in investment decisions.

The rest of this chapter is structured as follows. Section 10.2 presents the state-of-the-art on the design of recommendation systems for finance and motivation for a fuzzy-based approach. Then, Sect. 10.3 gives the introduction to recommender systems, the stock exchange simulation platform used by authors for the analysis, and the stock market terminology. Section 10.4 proposes the model and architecture of the recommended system. Later, the results of a use case is analyzed in Sects. 10.5 and 10.6 presents the lessons learned, conclusion and outlook of the study.

10.2 Literature Discussion

This section discusses the state-of-the-art on the design of recommendation systems for financial and banking services. In recent years, many researches about Recommender Systems (RSs) have been done to exploit the advantages of RS in the field of Banking and Finance. RSs have become more and more popular in the financial sector through applications and frameworks founded by researchers. Most of the applications are implemented based on non-personalized recommendations to improve the accuracy of predicting future trends because personalized recommendations require individual information of investors. This sensitive information is really difficult to achieve since customers do not want to share their personal information related to the investment budget. Another approach of RSs—is considered to provide recommendations in stock market—is case-based recommendation. These kinds of RSs use Case-Based Reasoning (CBR) to propose recommendations relied on the similar cases in the past [10].

The following are two representative applications of RSs in the financial domain.

Firstly, a RS framework is introduced by Musto and Semeraro in [5], this framework is based on the Case-Based Reasoning (CBR) in [3] and four different techniques—namely, Basic Raking, Greedy Diversification, Financial Confidence Value (FCV), and the combination of FCV and Greedy—to generate personalized portfolios. These portfolios are used by financial advisors to provide their clients investment proposals. The prototype of this framework is evaluated against 1172 real users and its result shows that the amount of profits in investment proposed by the framework overcomes those proposed by human advisors.

Secondly, an application of RSs in the financial domain is introduced by Gigli et al. in [2]. This application was developed by using three different RSs algorithms, namely Bayesian Personalized Ranking, Alternating Least Squares and Asset-Embedding—an adaptation of the Word2Vec algorithm—and its implementation has been done on a real dataset provided by the MPS bank. The results of this application shows that all three above-mentioned algorithms perform well on the real dataset. The Bayesian Personalized Ranking performs as the best algorithm in RSs for banking and financial services [2].

The above-mentioned application and framework were developed based on different techniques of RSs. However, none of them used the fuzzy-based approach. In this work, the authors present a fuzzy-based recommender system in the case "Börsenspiel for Swiss Universities (BSU)".

10.3 Theory

This section covers the concepts needed to understand the context of the case of study. Firstly, an overview of the recommender systems theory background is provided, in order to understand the elements and its characteristics. Secondly, the authors describe a brief introduction about the stock market simulation platform used in the analysis. Finally, a short walk-through in the stock market terminology is discussed in order to understand later the role that the stock characteristics play in the mappings with respect to our mathematical model.

10.3.1 Recommender Systems

Customers who visit a company's online store frequently see one or more recommendations of the kind "Customers who bought Item A also bought B" or "Customers who read a book in Finance also read these books in banking". Those recommendations are generated thanks to analytics tools that explore the behavioral patterns of customers who buy the same product [7].

There are different definitions of a recommender system (RS). In this paper, a RS is a software and techniques that provide suggestions about articles that should be shown to a particular visitor [4, 7].

A RS needs to know something about each user. Therefore, it maintains a user model which contains user profile data (especially user preferences), and remembers online activities of that user, in order to predict the articles that might be interesting for him. The way a RS collects this information depends on the particular Recommendation technique: user preferences can, for instance, be collected implicitly by monitoring user's behavior, or a RS might explicitly ask the visitor about his or her preferences. Moreover, it is important to collect additional information about each visitor's opinions and tastes [7].

The collected customer data are huge in volume and variety. The challenge of RS is to mine meaningful information to generate customized recommendations for each visitor.

10.3.1.1 Recommender System Typology

In the domain of recommender systems (RS), there are several categories, which classify the RS algorithms in terms of their source of information, techniques to obtain such information and the kind of recommendation provided. Depending on the characteristics involved in the different RS categories, it is possible to determine the application context (i.e., e-commerce, e-government, etc.). The different categories of RS are not only limited to the ones mentioned in this section, due to the possibility to combine them and add more characteristics, which eventually can not fit in these categories anymore, thus a new category is created. In our particular case, our FRS takes different elements of each of these categories without creating a new one. In order to understand our model and design in further sections a brief explanation of the RS typology is provided.

Collaborative Recommendation
With the collaborative type, the recommendations are addressed to customers who are willing to share similar interests (e.g., reading tastes, music listening preferences). As a consequence, every time a customer purchases a new product, this one is recommended to other users, too [7].

As an example, let's assume that customers X and Y have strong similarities in purchasing books. Let's now imagine that X recently bought a book not yet read by Y. A collaborative RS should propose this book to Y by filtering the most recommended one from a large set of books. This filtering happens as if customers X and Y cooperate to share their reading interests [7].

Collaborative RS are widely implemented in the context of e-Commerce where customers shop online. The advantage of this filtering technique is that the recommender system needs only the product id to make the right recommendation [4, 7].

Content-Based Recommendation
The techniques in this RS type are based on the availability of product specifications, which are, on the one hand, a user model and on the other hand a series of inputs that the user assigns a certain degree of importance/relevance to each product specification, such as product category (i.e., dictionary, travel guides, history), author, customer reviews [7].

A content-based RS must be capable to continuously update user models from user profile data in order to recommend relevant products to users having common interests or similar purchases [7].

A content-based recommendation system has two advantages [7]:

- It does not need large collection of data from large user groups to achieve relevant, accurate recommendation.
- New products are immediately candidates for recommendation to the users as soon as their specifications are made available to the RS.

Knowledge-Based Recommendation

There are market segments selling products characterized by a multi-year lifetime (e.g., PC, Smartphone, consumer electronics markets). The customers usually purchase these products once every one, or more years, in contrast with buyers of perishable products. Furthermore, customers of electronics products are not required to have advanced digital knowledge [7].

These market segments raise new challenges in the design of a recommender system [7]:

- Customers who buy a product once every one or more years do not have long purchase history for that product.
- As a consequence, a recommendation system must rely, not only on detailed knowledge about product technical specifications such as its performance, its compatibility with different software platforms, etc. But also it should be based on the knowledge and experience of users who review the product.

Systems that bring an answer to these challenges are called knowledge-based RS, subdivided into two categories: constraint-based RS and case-based RS.

As an example, let's consider an online store selling digital cameras. A constraint-based system must first acquire knowledge about cameras, such as resolution, weight, and price. A potential customer might search online for cameras with two explicit constraints: pixel count must be equal to or greater than 24 megapixels, brand must be Nikon. The RS should also evaluate these resolution and brand constraints with respect to the relative importance assigned by the customer in his profile to other camera specifications [4, 7].

Hybrid Recommendation

The combination of the previous types of recommender systems such as collaborative, content-based and knowledge-based results in a hybrid approach that might generate better, more precise recommendations to customers in specific circumstances. There are at least four possible combinations: *Collaborative and Content-Based*, *Collaborative and Knowledge-Based*, *Content-Based and Knowledge-Based*, and *Collaborative, Content-Based and knowledge-based.*

Intuitively, those combinations overcome the weaknesses of pure collaborative or pure content-based or pure knowledge-based RS. For example, community knowledge (collaborative RS) could be combined with product information (content-based RS) to design an enhanced RS, so-called hybrid.

In our research, the recommender system prototype was built as a hybrid RS since we generate recommendations based on both content and knowledge.

10.3.1.2 Recommendation System Properties

The basic properties of recommender system are: prediction accuracy, coverage, confidence, trust, novelty, serendipity, diversity, utility, risk, robustness, privacy, adaptivity, and scalability. Our FRS will attempt to satisfy some of these properties along with fuzziness properties. Indeed, for a given RS some properties might be more important than others (e.g., a designer could give less weight to prediction accuracy than diversity, risk, and privacy). He can vary the properties' weights to analyze their effects on the relevance of generated recommendations [7].

10.3.2 Introduction to Börsenspiel for Swiss Universities (BSU)

BSU[1] is a non-profit organisation (NPO) founded on 20 October 1991 by two students of the University of Fribourg, Andreas Hüchting and Harald K. Berg, under the name SEFU (Stock Exchange of Fribourg University). In 2015, the BSU president was David Chenaux, a student at the University of Fribourg. The statutes of BSU declare the following aims of this NPO in Article 5: (1) to gain an insight into stock markets, (2) to encourage economic thinking, (3) to bridge the gap between theory and practice, and (4) to gain experience with group dynamics through teamwork.

The BSU Simulation Game Rules

In 2015, each user must register to obtain an account and receive the initial fictitious amount of 1,000,000 CHF to start his game, and the registration is free. At the end of each day, the real-world stock market of SIX is replicated into BSU stock database for the simulation game. The aim of the simulator is to train the users to invest in different stocks. For example, by analyzing stock database, a user might decide to sell his stocks at the end of day if their prices are advantageous. Those players who perform best with highest return will be rewarded a CHF 50 gift from BSU's sponsoring companies. Players can buy or sell stocks at any time, but their transactions will be put in the pending status until 17:35 every day. At 17:35, those transactions will be executed at the fixed end-of-day rate (price). After a simulation period of 8 weeks, the winners are the top investors in the following categories: Performance and Risk.

[1] http://bsu.unifr.ch/.

10.3.3 Stock Market Terminology

This section explains the terminology used in BSU stock market game in this paper, as defined in Swiss Exchange online glossary [9]. Table 10.1 shows the most common definitions in terms of stocks characteristics.

Table 10.1 Stock market basic terminology

Name	Description
Category	Acronym of a stock market, e.g., SMI, DAX, CAC 40, and Structured Products
Interest	The return of a security, i.e., a stock in the BSU context
Performance	The variation of a stock price (increase/decrease) in a specific time interval, such as 1 week or whole game period (8 weeks)
Rate	The latest price of a stock in BSU simulation game
Risk level	Depends on the variation. If stock price variation is large, it means that a stock has high risk, and vice versa
Sharpe ratio	Automatically computed in BSU to evaluate the performance of investment based on risk-adjusted return. The formula is $$\frac{\textit{Average portfolio return} - \textit{Risk-free rate}}{\textit{Standard deviation of portfolio return}} \quad (10.1)$$
Stock high	The highest price of a stock during a given period, e.g., from 18/5/15 to 23/5/15
Stock low	The lowest price of a stock during a given period, e.g., from 18/5/15 to 23/5/15
Stock portfolio attitude	Each generated set of recommendations suggests three portfolio strategies corresponding to three attitudes; namely, Conservative, Explorer, and Adventurer. It is up to the investor to adopt an attitude and the corresponding portfolio strategy
Stock price	The amount of money an investor pays when he purchases a stock or the amount of money an investor receives when he sells a stock [1]
Variation	In this paper, the variation within a given period is a ratio computed by the recommender system using the following formula: $$\frac{\textit{Stock high} - \textit{Stock low}}{\textit{Stock price at the beginning of a given period}} \quad (10.2)$$

10.4 Fuzzy-Based Recommendation Model and Architecture

The design and implementation of a recommender system (RS) considers two main phases—namely, online and offline. In a first phase our fuzzy recommender was designed based on an offline evaluation from the BSU platform with a limited set of 20 stocks. This helped us to identify different types of elements that play main roles in the stock market and test our algorithm. The second phase consists of an online evaluation where the algorithm is implemented directly in the BSU platform. Thus, the algorithm has access to all the different stocks available and the user has the possibility to interact with the system.

This section presents the offline RS approach and in Sect. 10.5 the results of the online RS implementation are shown.

As part of the design and architecture modeling, it is important to define three main elements: taxonomy of the model, properties of the model, and requirements. These three elements are explained in subsequent sections.

10.4.1 Taxonomy of the Model

Table 10.2 shows the taxonomy of our Fuzzy RS,[2] in order to understand the context in which the recommender system can be applied.

The taxonomy table allows readers to understand the context in which the Fuzzy RS is applied. For example, it specifies the domain in which the RS is applied, its purpose, the scope of the recommendation in the personalization level, etc. This is helpful in order to have a detailed description of the of the Fuzzy RS.

Table 10.2 Taxonomy of the recommender system

Dimensions	Description
Domain	Stock market
Purpose	Recommend stocks to invest
Recommendation context	Listed companies on stock markets Switzerland, Germany, France
Personalization level	Individual student user or group as a whole (single account each)
Privacy and trustworthiness	Very high (it does not uses private information)
Interfaces	Direct online interaction with users in web environment
Recommendation algorithms	Fuzzy logic recommender

[2]Adapted from the course "Introduction to Recommender Systems" of Joseph A. Konstan, University of Minesota, United States.

10.4.2 Properties of the Model

Once the taxonomy is specified, the selection of the RS properties are defined, it is important to mention that it is up to the RS designers to decide which properties are covered by the RS. Ideally a good design should cover most of them, in order to have a greater impact on the recommendation; however, the more properties are covered, the more complex it is. In this particular case, we intended to cover the ones that can be linked directly to the stocks or investment instruments and specify the way to measure them.

Table 10.3 resumes the properties considered in the analysis of the stock and investment products in the BSU platform and their corresponding measurement technique.

10.4.3 Requirements

Once the taxonomy and properties have been defined, it is time to define the RS requisites, in other words, the key elements in which the recommendation should be based on. These elements will be considered by the RS and be able to make recommendations. The way to discover these requirements is analyzing the dynamic of the investor decisions in the platform. For instance the investment period, stock type, invested amount and currency are some basic key elements that investor considers in each investment in order to bet in a possible profit.

Table 10.3 Recommender system properties

Property	Measurement technique	Type(s)
Accuracy	Measure based on the comparison of the performance of the algorithm decisions and human expert investors	Quantitative
Coverage	Measured based on the number of stocks that a finance system can provide us	Quantitative
Trust	Based on user's feedbacks	Qualitative
Diversity	Diversity measure, relative diversity, precision-diversity Curve, Q-statistics, Set theoretic difference of recommendation lists	Quantitative
Risk-based	Automatic evaluation of risk based on daily, weekly variation of stock prices	Qualitative
Robustness	Prediction shift, average hit ratio, average rank	Quantitative
Privacy	High privacy since our algorithm does not rely in the user information	Qualitative/quantitative
Adaptivity	It is measured by the response of the algorithm based on the update rate of the stock information	Quantitative/qualitative
Scalability	It highly depends of the infrastructure to be implemented	Quantitative

Table 10.4 Requirements

Requirements	Description
Performance	The positive percentage behavior of the stock (the better the performance, the better the interest earnings in these stocks)
Short-term variation	The variation of the value of the stocks in the short term (1 week)
Investment	The amount of money invested in a period of time
Stock type	The stocks are from Switzerland, Germany or France
Reputation currency	Invest in CHF or Euro

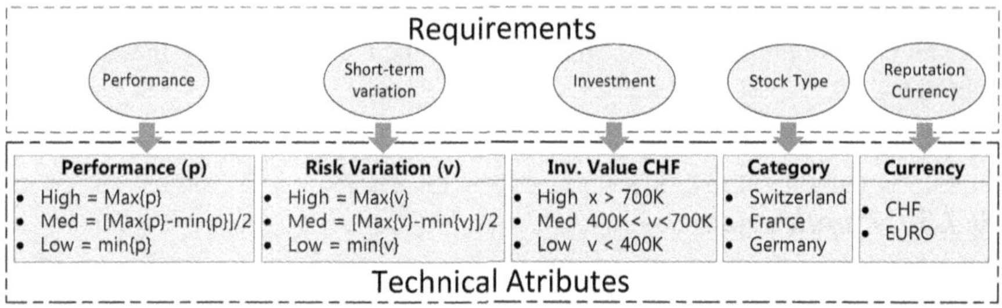

Fig. 10.1 Mapping between requirements and technical attributes

In our case, we considered 1 week of investment data from the platform and performed an one-time offline analysis on the behavior and parameters that the users considered to select their stocks. Then, their behavior and investment results allowed us to define five types of main requirements (Table 10.4).

10.4.4 Technical Attributes and Assumptions

Here is time to connect the previous defined requirements with their respective metric or technical attribute, which can later be changed in the algorithm to perform certain kind of simple classification or filtering. In Fig. 10.1 every requirement is mapped with a technical attribute, which help us to classify the stocks with five variables.

10.4.4.1 Fuzzy Sets and Membership Function

A fuzzy set in general terms without being strict in the mathematical definition, can be understood as uncertain sets, whose elements have degrees of membership. For example consider that our fuzzy set contains elements to describe the experience to say if the food in a restaurant was delicious or not. Each of the fuzzy set elements can contain for instance two values, one that says the level of being delicious and the other value to be terrible. Then a customer opinion about the food says that he enjoyed the dessert but the main course was not exquisite, it can be represented in a fuzzy variable that has 0.3 delicious experience and 0.7 non delicious experience, thus these elements in the fuzzy set are more adequate to be understood by humans. Fuzzy set elements are not black or white, they always have certain level or membership with respect certain parameters, in this case if the food was delicious or not [6].

Membership function for fuzzy sets is a generalization of the indicator function, which indicates if a value belongs to a set. In fuzzy logic, it represents the degree of truth, which are often confused with probabilities, although they are conceptually distinct, because fuzzy truth represents membership in vaguely defined sets, not likelihood of some event or condition. The Shape of the membership function used defines the fuzzy set and so the decision on which type to use is dependant on the purpose. Some examples of membership functions are: triangular, trapezoidal, Gaussian, etc. [6].

Last but not least, the threshold or line in the membership function, which establishes the frontier to decide when an element belongs or not in certain degree to a category, it is called alpha-cut [6]. A proper example is introduced later in the design of our FRS and it will be explained in more detail, for the moment it is important to keep in mind that it is just a frontier that decides if an element is classified in a category.

The membership function choice and alpha-cut are the subjective aspect of fuzzy logic and they allow the desired values to be interpreted appropriately [6].

In terms of fuzzy logic, there are not black or white terms, for instance lets consider to turn ON/OFF a light bulb, in strict terms it has only two states: ON or OFF. In the fuzzy world the bulb can be also in an intermediate state in which it is half ON and OFF, in other words, in medium light intensity. As a consequence one can define with more granularity as many intermediate sub-states between ON and OFF. This granularity and classification can be determined by a fix threshold or by a membership function. Therefore, the membership function can be seen as the rule that determines a category level of certain requirement.

In our case the membership function allows to provide a degree of membership of each stock in every requirement and map it into a fuzzy set with particular characteristics. In addition, the border parameter that determines the membership degree is in this case an alpha-cut, in other words, the alpha-cut determines the border between two states. For instance, considering the example of the bulb. If our alpha-cut, which has a value between 0 and 1, has a value of 0.6, it means that if a value lies exactly in the alpha cut, it means that the bulb state is turned ON in a 0.4 and OFF in 0.6 fraction. Thus the alpha-cut determines how strict the threshold between sub-states is and the shape of the membership function, provides the degree of belonging of the parameter.

In general, there are several membership functions in the fuzzy context (i.e., trapezoidal, singleton, Gaussian, triangular, etc.) in order to categorize, analyze products, services or processes. In our particular case we considered a triangular function with an alpha-cut parameter to fuzzily categorize the stock in what we are going to name fuzzy sets (see Fig. 10.2).

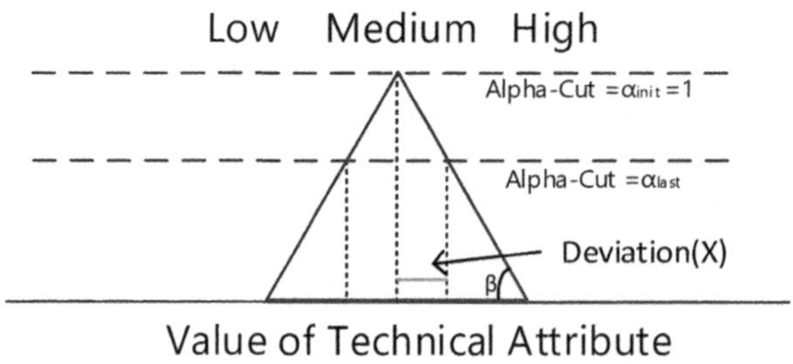

Fig. 10.2 Triangular membership function

In our case the way to determine the value of the deviation X in a triangle, is based on an assumption of a right triangle, where the tangent of an angle (i.e., β) is the ratio between the length of the opposite side and the length of the adjacent side. This definition gives rise Eq. (10.3) which is used to compute the deviation.

$$tan\beta = \frac{\alpha_{init}}{Max - Med} = \frac{1 - \alpha_{last}}{X} \Rightarrow X = \frac{(1 - \alpha_{last}) \times (Max - Med)}{\alpha_{init}} \tag{10.3}$$

The decision to choose a triangular shape as a membership function relies on the argument of faster implementation and the simplicity to implement [8]. Other types of membership functions can be considered as well but we let these improvements as part of the future work.

10.4.4.2 Fuzzy Filters

Once the membership function takes the data from the technical attributes, the categorization gives to the stock a certain degree of membership on a fuzzy set. The fuzzy set has three categories: *Performance Attitude*, *Risk Attitude*, and *User's Budget*. In addition, the stock by some information provided by the user can be filtered:

- **Countries list:** if the user prefers to invest in certain country (Germany, Switzerland or France or partially in some of them).
- **Reputation Currency:** Currency trust in case if the users prefers CHF or Euro.

10.4.5 Stock Portfolio

Once the fuzzy sets and filters have added a certain degree of characterization to every stock in a particular investment period, then the stocks are finally classified in three main Stock Portfolio categories that are shown to the user as a final result (Fig. 10.3).

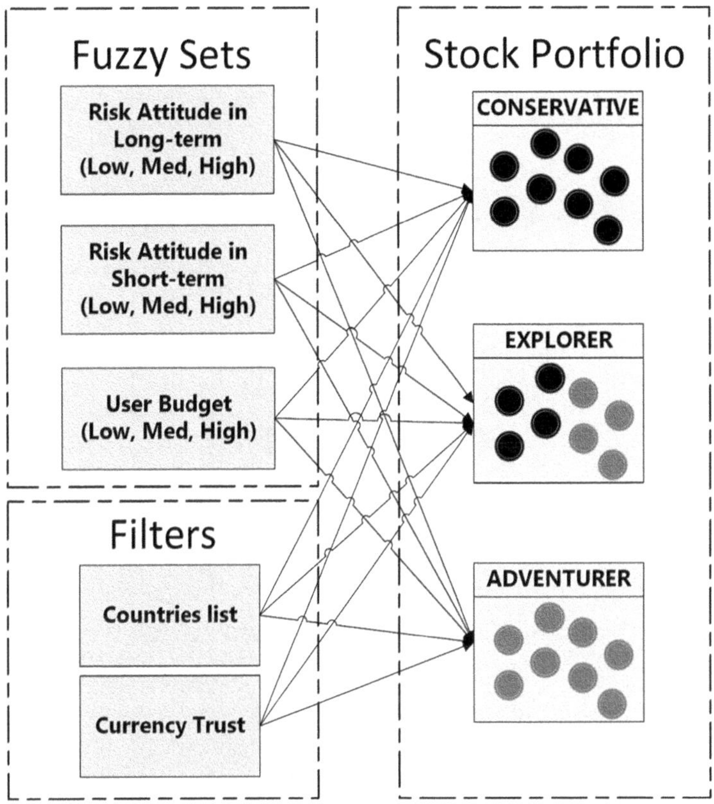

Fig. 10.3 Stock portfolio

The three main categories from the stock portfolio are conservative, explorer and adventurer. Each of these categories contains a set of subsets of stocks where the user according to the investment strategy preferences receives a set of recommended stocks to invest. In other words, the stock categories are constructed based on the fuzzy sets. In our case the conservative, explorer and adventurer are formed by combinations of low, medium and high classified stocks in each of the fuzzy sets categories.

The next subsections describe more in detail the stock portfolio categories and the requirements that the stocks have to fulfill to belong into a specific stock portfolio. We present it in a decision flow diagram per stock portfolio category.

In all cases regardless the user preferences or the level of risk to invest, the algorithm decides always to find the best performance in the stocks, which is reflected in the fuzzy set Performance Attitude and considers only medium and high stocks that fall in these categories. This slight criterion reduces significantly the risk of losing money as we would discuss later in the first test results.

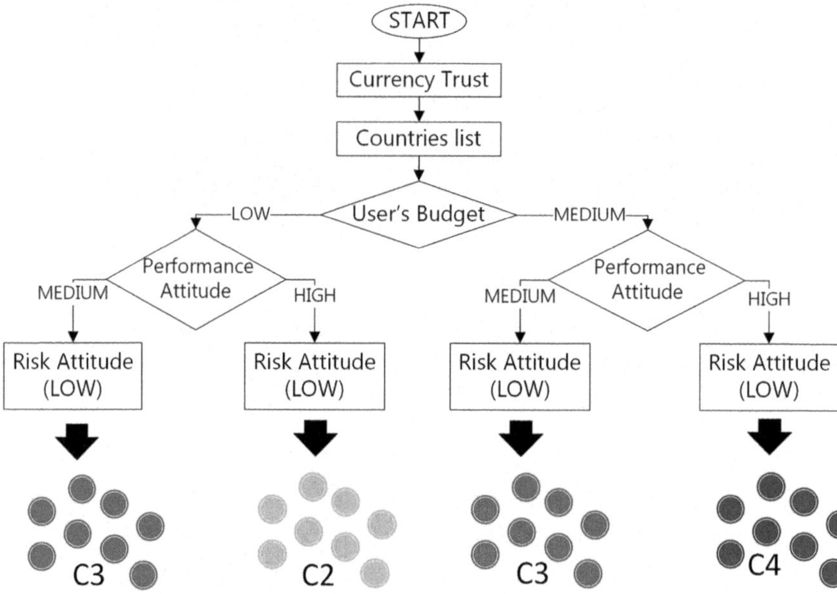

Fig. 10.4 Conservative profile

10.4.5.1 Conservative Profile

The conservative profile is designed for the investors who do not want to take too much risk and count with a low or medium budget to invest. Moreover, it contains stocks that have the characteristic low in all the fuzzy sets. Figure 10.4 shows the flow diagram of the conservative profile. The outputs are four subclasses of stocks inside of this category that would be suggested to the user.

10.4.5.2 Explorer Profile

The explorer profile is designed for the investors who are 50% conservative and 50% risk-takers. Moreover, it contains stocks that have the characteristics low and medium in all the fuzzy sets. Thus, the RS must have all the combinations between low and medium to be able to recommend stocks in those categories.

These investors count with a high or medium budget to invest. Figure 10.5 shows the flow diagram of the explorer profile. The outputs are eight subclasses of stocks inside of this category that would be suggested to the user.

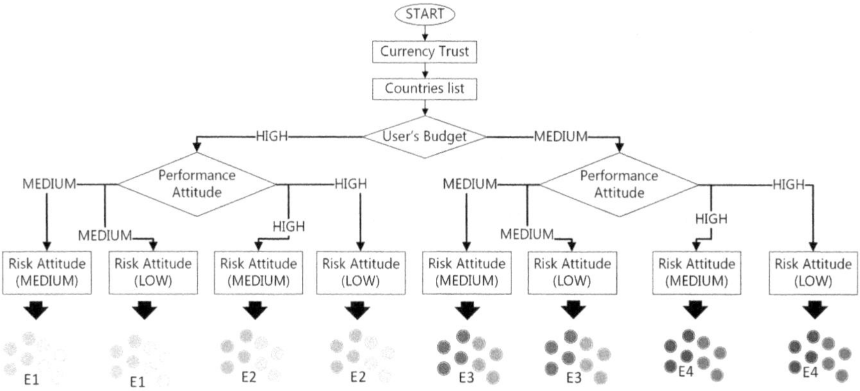

Fig. 10.5 Explorer profile

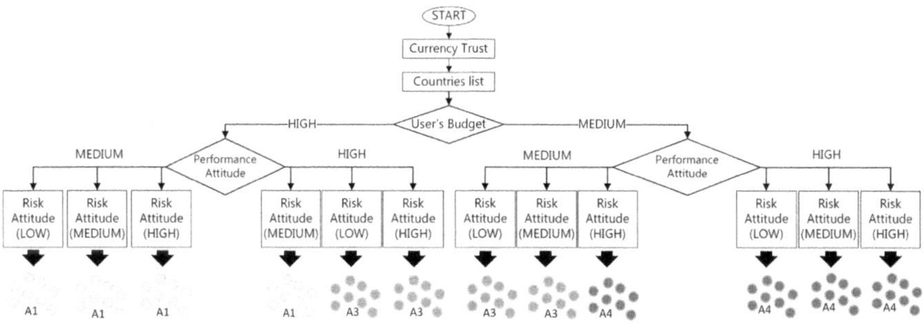

Fig. 10.6 Adventurer profile

10.4.5.3 Adventurer Profile

The adventurer profile is designed for the investors who are totally risk takers, these investors forgot to be afraid of losing money long time ago and they count with a high or medium budget to invest. Moreover, it contains stocks that have as characteristic low, high and medium in all the fuzzy sets. Figure 10.6 shows the flow diagram of the adventurer profile. The outputs are four subclasses of stocks inside of these categories that would be suggested to the user.

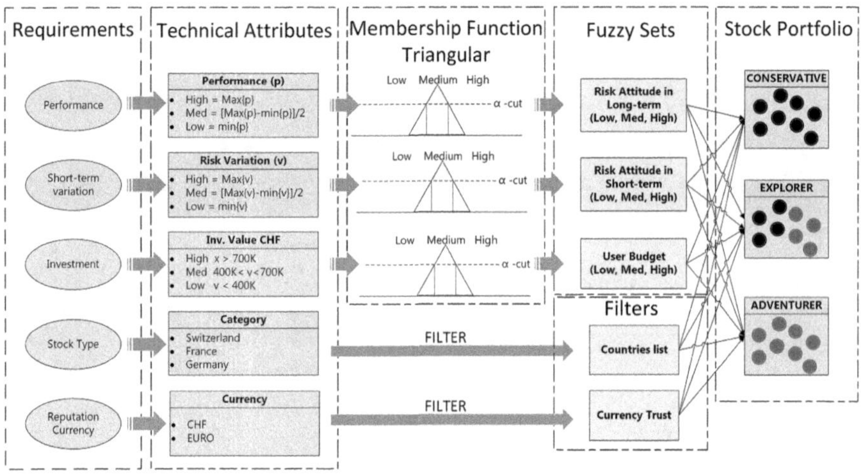

Fig. 10.7 Fuzzy recommender architecture

As we can observe in the way that users are more flexible in taking more risk in their investment behavior, the algorithm would be more dynamic, recommending more stocks. At the end the stocks classified as low are always included in all of the portfolios on purpose, in order to compensate the loses to a certain degree in all of the profiles.

10.4.6 Model Design

After reviewing all the different elements that are involved in the algorithm model, we can have an overview of the entire architecture of the fuzzy recommender algorithm in Fig. 10.7.

10.4.7 Decision Criteria

The most critical part of the algorithm is to find a match between the parameters (fuzzy sets and filters) with the stock portfolio. Figure 10.8 shows that the fuzzy sets and filters form a vector, if the membership function could classify correctly the parameters then all the vector contains zeros, which means that the stock has a match and can be classified directly to the three main Stock Portfolios. Nevertheless, there

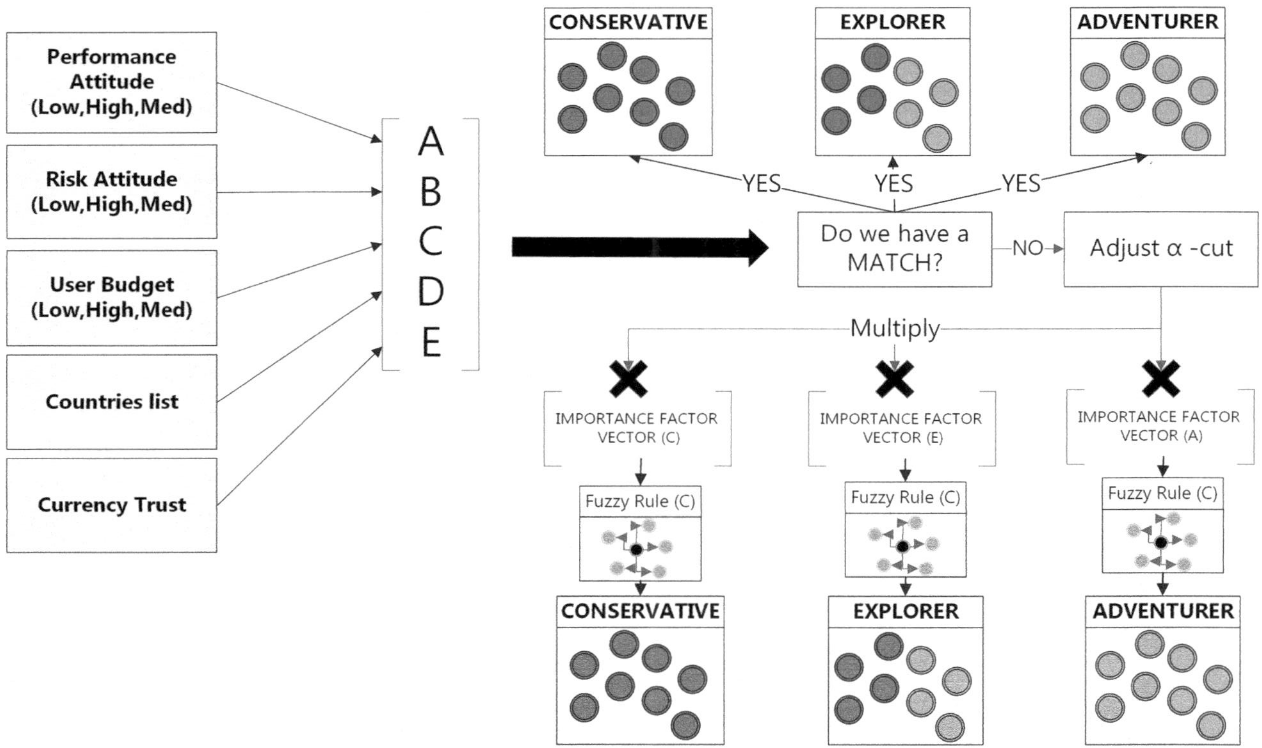

Fig. 10.8 Fuzzy algorithm decision criteria

are some stocks that cannot be classified properly because they cannot match 100% by the membership function, so if there is no match for these stocks the membership function will be computed again with a lower value of alpha cut until it finds a match in the criteria and then the vector would contain the value of the difference between the original and the new alpha cut (Fig. 10.8).

For instance let's consider that a stock in the Performance Attitude cannot be matched, and then the value of A in the vector would have the difference between two alpha cuts instead of zero. Then we multiply the vector of the stock for an importance factor vector that takes into account per category the most important values considered for the stock. Finally, if the importance vector did not make the whole vector of the stock zero (close to zero), then a fuzzy rule per category is taken into account. As an example of fuzzy rule for the conservative category would be that the value of A is zero if the value is lower than 0.1. The purpose of these different criteria is that the algorithm is able to classify all the stocks that are considered in the analysis and find a best suitable category where it belongs.

10.5 Use Case: BSU Simulation Game Assisted by Fuzzy Recommender System Prototype

This section presents the results of our RS prediction algorithm. We will compare the earnings predicted by our FRS with the earnings of the top three investors who won a BSU game in 2015. BSU presents approximately 600 stocks from four stock markets: SMI (Swiss Market Index), DAX (Deutscher Aktienindex—German stock index), CAC 40 (Cotation Assistée en Continu—a benchmark French stock market index), and Structured Products (a market-linked investment).

For offline RS test purposes, our fuzzy recommender processes only stocks from 20 listed companies in SMI. The resulting recommendations correspond to three attitudes: conservative, explorer and adventurer.

10.5.1 Fuzzy Recommender Pre-Processing Results

Our fuzzy algorithm requires input for the whole game period (8 weeks). To this end, we extract once a week the profile of each stock which consists of the following data:

- category
- company name
- short-term (last week): stock price on Monday, stock price on Friday

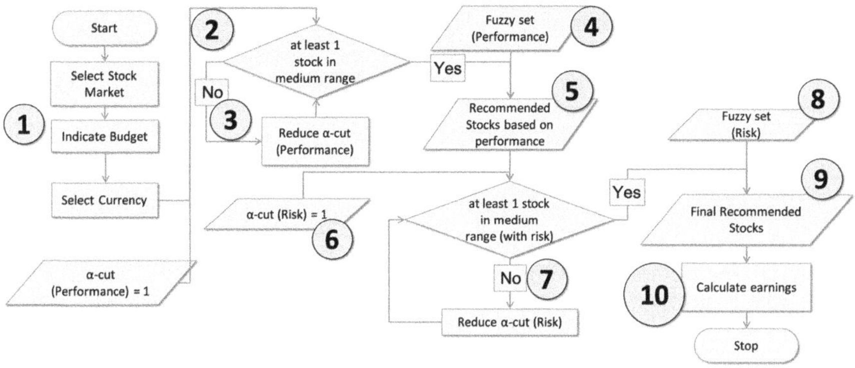

Fig. 10.9 Fuzzy algorithm decision criteria

- short-term (last week): stock high, stock low
- long-term (whole game period): stock price at start of game, stock price at end of game
- long-term (whole game period): stock high, stock low

10.5.2 Fuzzy Recommender Algorithm

Figure 10.9 shows the flowchart of our Fuzzy Recommender Algorithm. The algorithm steps are explained below, along with the circled digits as needed:

10.5.3 Findings and Interpretation of Results

After running our algorithm on BSU platform from each of the investment scenarios, the FRS performed among the top three investors by investing in its recommended portfolios and its earnings were positive in all the cases. This section presents in details the different findings and results from each of the investment scenarios by round periods. The periods of investment were measured in terms of weeks.

Algorithm 1 Fuzzy recommender algorithm

1. Select Stock Market (SMI, DAX, CAC 40) → Indicate the amount of money that the investor would like to invest → Select the currency (EURO for DAX and CAC 40, CHF for SMI) → initialize α-cut of performance to 1.
2. Is there at least one stock in the medium range of membership function Performance?
   - If YES, Go to step 4
   - If NO, Go to step 3
3. Reduce α-cut of performance by 0.1 → Go to step 2
4. Assign Low, Medium, and High to Performance fuzzy set according to one of possible attitudes (Conservative, Explorer, Adventurer). For example, assign Medium and High to Performance fuzzy set of Conservative attitude.
5. Output recommendations on stocks to buy based on performance.
6. Initialize α-cut of risk to 1
7. Is there at least one stock in the medium range of membership function Risk?
   - If YES, Go to step 9
   - If NO, Go to step 8
8. Reduce α-cut of risk by 0.1 → Go to step 7
9. Assign Low, Medium, and High to Risk fuzzy set according to one of possible attitudes (Conservative, Explorer, Adventurer). For example, assign Low to Risk fuzzy set of Conservative attitude.
10. Initialize importance vector for a given attitude → Calculate the distance vector → compute the vector result of the multiplication between importance vector and distance vector → Apply fuzzy if needed
11. Output final recommendations on stocks to buy based on both performance and risk.
12. Compute the earnings for a given attitude.

10.5.3.1 Scenario A: Comparison Between FRS Earnings and Players' Earnings After 1 Week

Figure 10.10 shows the top three investors and the FRS at the first week or round of investments. In both cases every user started with a virtual capital of one million francs. In the case of the FRS, it also shows the results of the investment based on the three strategies: conservative, explorer and adventurer.

Considering the results of the first week it is possible to notice the following observations:

- Users who are conservative or explorer have almost the same level of earnings
- Conservative and Explorer earnings are positive, but lower than the earnings of the top three users.
- Regarding the scenario of the Adventurer, his earnings are better than the second top user's earnings, but still lower than the first top user. It is important to emphasize that in the BSU platform, the strategy differs week by week for the users, and they may select to be conservative in a certain point or take more risks. In our FRS the algorithm keeps the same strategy during 1 week.

	Investment Amount		Earnings (CHF)
Human Decisions	CHF 1'000'000	BSU Top 1	225'669
		BSU Top 2	156'713
		BSU Top 3	151'253
Fuzzy Recommender System Predictions	CHF 1'000'000	FRS Conservative: max earning	41'557
		FRS Explorer: max earning	41'557
		FRS Adventurer: max earning	167'401

Fig. 10.10 Scenario A

	Investment Amount		Earnings (CHF)
Human Decisions	CHF 1'000'000	BSU Top 1	1'805'350
		BSU Top 2	1'253'702
		BSU Top 3	1'201'207
Fuzzy Recommender System Predictions	CHF 1'000'000	FRS Conservative: max earning	332'457
		FRS Explorer: max earning	332'457
		FRS Adventurer: max earning	1'339'207

Fig. 10.11 Scenario B

The question is whether better results can be obtained if the algorithm changes the strategy every period (1 day, 2 days, 1 week, ...).

10.5.3.2 Scenario B: Comparison Between FRS Earnings and Players' Earnings After 8 Weeks

In this scenario, we consider again the top three winners in the platform and compare their earnings with three users assisted by our FRS during the same game period (8 weeks). Figure 10.11 shows the earnings of the same amount invested by the best users and the fuzzy algorithm.

The Conservative and Explorer investors earnings still have a positive net income after 8 weeks, and the Adventurer is still better than the second winner with 1,339,207 Swiss Francs (CHF) (as shown in Fig. 10.11). Unfortunately, our FRS with the Adventurer attitude cannot surpass the best talented trader of BSU. This weakness comes from our FRS which is still the bottom line of the prototype in its infancy. We need to scour other good investment criteria, enhance our membership function, and analyze stock prices daily to get better results. Another reason for the weakness is that, in this paper, we considered only one stock portfolio during a period of 1 week or the whole game. If investors who use our FRS change their stock portfolio every week, or possibly each day, they might get better results.

10.6 Lessons Learned, Conclusion and Future Work

The project to implement the recommender system raised several challenges and satisfactions. On the side of the design of the recommender system, it took us into a trip in the finance field in order to understand concepts, dynamics and the rules of the stock market. Moreover, different trade-offs and variables to model took us some time to understand their real value for our recommender system.

The coding implementation and integration of the algorithm into the BSU platform was an interesting and challenging part, which involved to refactor JAVA code in the back-end, PHP and Javascript on the front-end.

It was an ambitious case study, which included not only the design, theory and documentation of the recommender system algorithm but also the integration in a live stock exchange platform that was running and used by several users at the same time. Thus, deploying and debugging code were also important skills developed during the project.

The field of fuzzy recommender systems is highly active and in constant evolution. Unfortunately, concerning stock recommender systems, the research literature is not yet abundant. The authorization of the BSU platform to implement our algorithm in their system, allowed us to perform our research. With his permission, we implemented not only the theoretical framework behind a Fuzzy Recommender Algorithm, but also were able to test and improve our algorithm.

The results of the different scenarios provided in this research come from a basic prototype implementing a solid algorithm architecture that might be applied in different stock market platforms. Based on the earnings comparison and results between the top users and analytics of our fuzzy algorithm, we can say that the algorithm behaves with promising results to generate revenue. The web-oriented prototype is the second version of our algorithm, who considers three investment attitudes. Investors can adopt one of them to make their decisions.

Finally, our fuzzy model is still improving to adapt to other stock characteristics and the next points are recommended for future work.

10.6.1 Formal Evaluation of Properties

The recommender system properties that characterize our proposed fuzzy algorithm need to be measured quantitatively. In order to evaluate the properties, a series of tests with different users are needed in order to obtain useful information from the investors that allow us to measure not only the accuracy of the algorithm but also other important properties as trust and risk for instance. The formal quantitative measurement of properties for our algorithm is a vital part for future analysis and specific improvements.

10.6.2 Different Membership Functions

For the purposes of the Excel version of the algorithm, the first version presented in this document considers a triangular membership function. Our future versions of the algorithm might consider new forms of membership functions such as

Triangular, Trapezoidal, Gaussian and Generalized Bell, that can offer better recommendations to financial analysts and increase the accuracy of our algorithm.

10.6.3 New Criteria for Variation Calculation

The current algorithm considers the variation of the stocks per week or a period of 8 weeks giving good results. Nevertheless, in order to look for improvements in terms of accuracy, new criteria to select shorter variation periods should be considered. For instance measuring the variation of the stock per day or twice a day could bring more accurate recommendations.

References

1. Ang, C. S. (2015). *Analyzing financial data and implementing financial models using R*. Basel: Springer.
2. Gigli, A., Lillo, F., & Regoli, D. (2017). Recommender systems for banking and financial services. In *Proceedings of RecSys 2017 Posters, Como, Italy*, August 27–31 (Vol. 1905).
3. Lorenzi, F., & Ricci, F. (2005). Case-based recommender systems: A unifying view. In: *Intelligent techniques for web personalization* (pp. 89–133). Berlin: Springer.
4. Mancera, J. A., & Bosshard, P. (2015). *The impact of recommender systems on business and customers in electronic markets*. Electronic Business Seminar, University of Fribourg.
5. Musto, C., & Semeraro, G. (2015). Case-based recommender systems for personalized finance advisory. In *CEUR Workshop Proceedings* (Vol. 1349, pp. 35–36).
6. Novak, V. (1989). *Fuzzy sets and their applications*. Bristol: Techno House.
7. Ricci, F., Rokach, L., Shapira, B., & Kantor, P. (Eds.). (2011). *Recommender systems handbook*. New York: Springer.
8. Schuetze, R. (2013). Intuitionistic fuzzy component failure impact analysis (IFCFIA). *Notes Intuitionistic Fuzzy Set, 19*(3), 62–72.
9. SIX Swiss Exchange Ltd: Glossary. Retrieved May 1, 2015 from http://www.six-swiss-exchange.com/knowhow/glossary.html
10. Zibriczky, D. (2016). Recommender systems meet finance: A literature review. In *International Workshop on Personalization and Recommender Systems in Financial Services*.

Printed by Printforce, the Netherlands